A Guide to Energy Management in Buildings

This new edition of *A Guide to Energy Management in Buildings* begins by asking why we need to control energy use in buildings and proceeds to discuss how the energy consumption of a building can be assessed or estimated through an energy audit. It then details a range of interventions to reduce energy use and outlines methods of assessing the cost-effectiveness of such measures.

Topics covered include:

- where and how energy is used in buildings
- energy audits
- measuring and monitoring energy use
- techniques for reducing energy use in buildings
- legislative issues.

And new in this edition:

- the cooling of buildings
- fuel costs and smart metering
- education and professional recognition.

It provides a template for instigating the energy-management process within an organization, as well as guidance on management issues such as employee motivation, and gives practical details on how to carry the process through. This book should appeal to building and facilities managers and also to students of energy management modules in FE and HE courses.

Douglas J. Harris is an assistant professor in the Centre for Excellence in Sustainable Building Design at Heriot-Watt University, UK. He carries out research and teaching on energy use in buildings.

A Guide to Energy Management in Buildings

Second edition

Douglas J. Harris

Routledge
Taylor & Francis Group

LONDON AND NEW YORK

Second edition published 2017
by Routledge
2 Park Square, Milton Park, Abingdon, Oxon, OX14 4RN

and by Routledge
711 Third Avenue, New York, NY 10017

Routledge is an imprint of the Taylor & Francis Group, an informa business

First edition published by Spon Press 2012

British Library Cataloguing in Publication Data
A catalogue record for this book is available from the British Library

Library of Congress Cataloging in Publication Data
Names: Harris, Douglas, 1951- author.
Title: A guide to energy management in buildings / Douglas Harris.
Description: Second edition. | New York, NY : Routledge, 2017. |
Includes bibliographical references and index.
Identifiers: LCCN 2016025016 | ISBN 978-1-138-12068-6
(hardback : alk. paper) | ISBN 978-1-138-12069-3 (pbk. : alk. paper) |
ISBN 978-1-315-65157-6 (ebook : alk. paper)
Subjects: LCSH: Buildings—Energy conservation. | Building management.
Classification: LCC TJ163.5.B84 H38 2017 | DDC 658.2/6—dc23
LC record available at https://lccn.loc.gov/2016025016

ISBN: 978-1-138-12068-6 (hbk)
ISBN: 978-1-138-12069-3 (pbk)
ISBN: 978-1-315-65157-6 (ebk)

Typeset in Sabon
by FiSH Books Ltd, Enfield

MIX
Paper from
responsible sources
FSC   FSC® C013056
www.fsc.org

Printed and bound in Great Britain by
TJ International Ltd, Padstow, Cornwall

Contents

Illustrations

Figures

Tables

1 Background

The energy problem

The importance of energy to a country's economy and the impact of energy use on the environment are matters which have received vastly increased attention in the last few years. The problem is often stated as an energy 'trilemma', which is defined as the need to

- improve security of supply
- reduce energy costs to consumers and businesses
- reduce carbon emissions to minimise the increase in global average surface temperature.

Climate change is now widely acknowledged to be a product of man's profligate use of fossil fuels, and it is well known that there is an urgent need to cut our fossil fuel consumption substantially over the next few years. Approximately 50 per cent of our energy use, and carbon dioxide (CO_2) emissions into the atmosphere, are from the use of energy for heating, cooling and lighting buildings; 25–30 per cent is used in transport, while the remainder is used for industrial processing. At the Kyoto summit in December 1997, the UK made a voluntary commitment to reduce CO_2 emissions by 20 per cent by the year 2010, and other countries made similar commitments to reduce their use of fossil fuels. While some countries made substantial cuts and exceeded their targets, others failed to meet their commitments; in emerging economies such as China, there have been sharp increases in emissions in recent years. More ambitious recent targets require a reduction of 60 per cent in carbon emissions by 2050, and the 2015 Paris Agreement heralds a new phase in global action to tackle climate change.

In addition to global treaties regarding carbon emissions, building regulations and codes of practice in the UK and many other countries are becoming increasingly strict concerning energy use. In Scotland, for example, the latest building regulations (2015) have been amended to produce a 43 percent reduction in CO_2 emissions for newly constructed domestic buildings relative to 2010 figures.

As buildings account for about 32 percent of the world's energy use (nearer 40 percent of primary energy use if embodied energy for construction is included), it is clear that reducing energy consumption in buildings can make a major contribution to lessening carbon emissions to a meaningful degree, and good management of energy use in buildings is acknowledged as an important aspect of sustainable development.

Sustainability

Minimising energy use in buildings is an important aspect of sustainability. In the Building Research Establishment Environmental Assessment Method (BREEAM) of assessing the environmental impact of buildings, having low carbon emissions is important in achieving a high sustainability rating for a building. Much BREEAM activity is directed at new-build and the BREEAM process can help steer design teams towards more sustainable solutions, but consideration of some of the issues fundamental to BREEAM and other assessment methods used around the world can also guide us towards better approaches to the refurbishment of existing buildings. A notable example is material use: assessment methods give credit for reusing and recycling materials, since they result in lower raw-material use and reduced embodied energy. Analysis of the embodied energy in a building may also help us decide whether to retain a building or to replace it altogether. The embodied energy of a large building may be substantial, and it is often better to keep the frame of a building and refurbish it in a low-energy way, than to demolish it and replace it with a new one. Valuable architectural features and the significance of the building as part of the urban fabric are further qualities that may argue for its retention, although they are harder to quantify than embodied energy.

The importance of this is reflected in the BREEAM environmental assessment method for buildings, of which energy consumption is a major constituent: over 30 per cent of the credits in this system are related to energy efficiency. It has been estimated that office buildings in the UK alone are losing £7 million a day through wasted energy. Improved standards are continually being required of new buildings – changes to Building Regulations Part L in 2010 meant that carbon emissions had to be cut by 25 per cent compared with 2006 levels. Carbon reduction is no longer something that is just talked about, it has become an integral part of the building manager's job, and people are now scrutinizing carefully their energy consumption and carbon emissions, at home and at work. The changes required to ensure continued improvements in this area demand a huge change in our attitude towards energy use and generation methods, and even with improvements in demand management we will need to generate 30–40 per cent of energy from renewable sources – amounting to an increase of about 1200 MW annually.

Reducing the impact of climate change by lessening our reliance on fossil fuels is only one of the reasons for wanting to lower our consumption of energy in buildings. Although global economic growth has not been even, and indeed in some places has been negative, in recent decades countries such as China have been expanding their economies rapidly, and with this expansion there has been an increased demand for energy, much of it provided by fossil fuels such as oil, coal and gas. This brings with it another aspect of the energy problem, that of continuity of supply of fossil fuel; fuel reserves are dwindling, and many observers estimate that oil production is already past its peak. If demand continues to soar while supply falls, prices will inevitably escalate at a rapid rate, leading to severe economic and social problems. Thus, for economic and environmental reasons, reducing energy use in buildings is imperative. Prior to a continued fall in oil prices in the last year or two, increases in oil and gas prices raised public awareness about the cost of energy. In the UK gas price increases of 25 per cent in less than a year led both domestic and business users to place a greater emphasis on energy efficiency, but during the writing of this book there was a period of increased oil production in Saudi Arabia and other Gulf states, which resulted in significant reductions in the oil price. Although there will always be periods of fluctuating oil prices, the long-term trajectory must inevitably be upwards. Many countries rely for their fuel supplies on increasingly expensive imports, therefore the need to reduce this reliance is a further driving force behind energy management. The impact of current legislation on energy efficiency, the use of new materials and techniques for utilizing renewable energy, and the adoption of more energy-conscious design, mean that most new buildings will consume much less energy per square metre than the majority of those already existing. Initiatives such as the Code for Sustainable Homes and the Passivhaus standard help guide architects towards low-energy solutions, although the UK Government came in for much criticism when it scrapped plans to have all new housing zero carbon by 2016.

The long-term effects of climate change will result in slightly lowered heating loads and increased cooling loads, but these are unlikely to have a significant effect on energy use. Energy consumption in existing buildings is the most important factor, as in most developed countries the rate of replacement of old buildings with new ones is very low (about 1 per cent per year) and the time scale for substantial reductions in CO_2 emissions is therefore very long (see Figure 1.1). The use of passive techniques such as greater thermal mass and better use of solar gains, which help to keep consumption down in many new buildings, is only applicable to a fraction of older buildings; in order to achieve a more rapid reduction in CO_2 levels, lowering energy consumption in existing buildings is consequently of major importance. Governments worldwide are instigating measures to improve sustainability throughout all areas of economic activity, and energy efficiency in buildings is an essential element of all these initiatives.

Against this background, the purpose of energy management is to analyse where energy is wasted in buildings and identify cost-effective

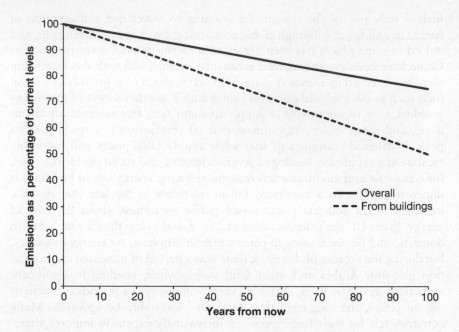

Figure 1.1 Effect on CO$_2$ emissions if every new building emitted at only 50 per cent of current levels without reducing energy use in existing buildings

solutions. While many techniques can be applied to both new and existing buildings, the contents of this book are aimed mainly at improving buildings already in use.

Energy use in buildings

In some respects the statistics paint a fairly gloomy picture, but they do highlight the need for urgency in applying carbon-reduction measures. Energy management as described in this book is largely concerned with demand-side management: that is, reducing the demand for energy in existing buildings by improving the fabric, plant and management of the building. While the use of renewable sources of energy such as photovoltaics (PV) helps to supply that demand with reduced carbon emissions, the focus here is on reducing demand at the outset. Once the demand is lowered, then more carbon-efficient sources of supply can be considered.

Domestic

Within developed countries in Western Europe, between 30 and 60 per cent of primary energy is used in buildings, chiefly to provide heating, cooling and lighting. In the UK, approximately half of this is used in housing. The distribution of energy use within different building types varies greatly. In housing, approximately 58 per cent of delivered energy is used for space heating, and space and water heating together account for 82 per cent of consumption. Since 1970, energy use for space heating has risen by 24 per cent, for water heating by 15 per cent, and for lighting and appliances by 15 per cent. Over the same period energy use for cooking has fallen by 16 per cent, and overall electricity consumption has risen by 63 per cent. This rise in electricity consumption is partly due to the vast increase in the use of home computers, electronic games and other entertainments.

Some of these changes are driven by changes in lifestyle and demographics – the overall number of households has risen sharply in the last twenty years, largely as a result of an increase in the number of people living alone. Central heating, which was used in only one third of houses in 1970, is now installed in 90 per cent of dwellings. There has also been a significant change in the type of housing built in the past twenty years: more one and two-bedroom flats, for example, to meet the changes in family structure. In spite of the large number of houses built in recent years, just under 40 per cent of the UK housing stock was built before 1945, and roughly 15 per cent built since 1990. Much of the older housing is unimproved in terms of energy efficiency. The greater use of the central heating has resulted in higher internal temperatures – average 18°C in 2000 as opposed to 13°C in 1970, and homeowners now tend to heat more rooms in their houses. The effect of this on national energy consumption is significant, particularly when one remembers that a 1°C rise in room temperature equates to about 7 per cent increase in energy consumption. It may be claimed that it is at least partly due to government initiatives that approximately 90 per cent is houses have loft insulation and over 60 per cent have some or all of their windows double-glazed. Of the 69 per cent of dwellings having cavity walls, about 50 per cent of them are now insulated.

Overall energy use per household has been falling steadily over the last few years as a result of improved thermal performance and improved efficiency of appliances such as washing machines, however energy use on computing and home entertainment has increased. The average home energy use per year is approximately 14,000 kWh of gas and 4,000 kWh of electricity after adjusting for ambient temperature, costing £630 and £420 respectively. Consumption in kWh is split roughly as follows:

- space heating   58 per cent
- water heating   24 per cent
- other electricity   14.6 per cent
- lighting   3.4 per cent

It has been estimated that it would cost the average homeowner £2000 to reduce their energy consumption by 25 per cent. On the basis of the above figures, this would produce an attractive payback period of just over two and a half years. Most of the techniques proposed rely on reducing the load (e.g. by adding insulation, etc.) and using more efficient electrical appliances from conventional mains supplies, but the use of renewables is increasing, leading to falls in future emissions. By April 2016 there was approximately 9213 MW of installed photovoltaic (PV) power in the UK spread over 867,867 installations, this number having risen rapidly since the introduction of feed-in tariffs in 2010.

Non-domestic

Non-domestic buildings cover a wide range of building types and uses. The most common are offices, and there is considerable documented evidence of their energy consumption, in the UK at least, but other building types produce substantial emissions. Hospitals are the largest emissions producers in the UK, with an average of 4089 tonnes of CO_2 per year, while prisons are second with 2849 tonnes. Typical figures for offices are shown in Table 1.1. As sizes vary greatly, typical consumption figures are given in kWh/m^2 of treated or usable floor area.

After lighting, computers and monitors have the highest energy consumption in the office. Studies show that about half are left on overnight and at weekends – about 75 per cent of the hours in the week. Simply turning them off at night can result in substantial savings.

Aims and objectives of energy management

Energy management in buildings is concerned with maximizing the use of energy resources while providing the desired environmental conditions and

Table 1.1 Typical office energy consumption and carbon emissions in the UK

Naturally ventilated cellular	*Energy kWh/m^2*	*CO_2 kg/m^2*
Heating and hot water	160	9
Cooling	0	0
Lighting	25	4
Office equipment	30	6
Total	*215*	*19*

Air-conditioned		
Heating and hot water	180	8
Cooling	30	4.5
Lighting	50	8
Office equipment	30	6
Total	*290*	*26.5*

services inside the building at the least cost. Energy management can also be applied to industrial processes, particularly those requiring the use of steam or heat, such as textile manufacture and steel making, but this book is concerned only with the use of energy in buildings to provide thermal, visual and acoustical comfort for the occupants. The range of building types that an energy professional may be required to examine includes homes, offices, schools, universities, hospitals, retail premises, sports centres, cinemas, theatres, restaurants, and other commercial premises.

The key aims of energy management are:

- minimizing energy consumption
- optimizing size of plant
- maximizing energy efficiency
- minimizing energy waste
- reducing carbon emissions
- reducing costs.

Energy management involves the measurement or estimation of the energy consumption of a building, the provision of recommendations for improvement, and the development of a strategy for maintaining continuous improvements in the energy performance. Typical techniques for reducing energy consumption will be discussed in detail later. They include:

- use of thermal insulation
- recovery of waste heat
- combined heat and power (CHP)
- the use of more effective heating and cooling systems
- the use of natural methods of ventilation and cooling
- better use of controls for heating, cooling and lighting
- a strategic approach to managing and reporting energy use.

Energy costs

The running costs of many organizations tend to be dominated by the cost of labour, with energy accounting for only 5–10 per cent. When savings are needed it is much easier simply to reduce staff numbers – which also reduces a company's capability. If the company's total energy bill is considered, however, the amounts involved can be considerable, and for energy-intensive industries with a small workforce, such as the Scotch whisky industry, energy may represent up to 50 per cent of the running costs. Furthermore, sudden increases in fuel prices can cause serious cash flow problems for small and medium-sized businesses. Energy efficiency struggles under the handicap that energy costs are often 'invisible', and it does not create an income, and when energy costs are reduced there is no visible increase in the income stream, only an estimate of what would have been spent if energy-saving measures had not been introduced. Clients need to be assured that they are getting value for their investments, and it is essential to be able to supply them with detailed figures for energy and cost savings.

Example

Consider a 25-year-old air-conditioned office building situated in the UK:

100 m × 20 m floor plan having three storeys, with a total annual energy consumption of 300 kWh/m^2 (i.e. a fairly inefficient building, but not atypical).

Total floor area = 6,000 m^2.

Total annual energy consumption = 100 × 20 × 3 × 300
                                  = 1,800,000 kWh per year.

Cost at average fuel price 10p/kWh   = £180,000 per year
or £30 per sq. metre per year.

With good energy management, 10–30 per cent per year could be saved, i.e. £18,000–£60,000. Individual cases may show larger savings for a relatively small outlay.

If we consider the total amount spent on energy by an organization in a year, then we can see that energy management can be worthwhile, and many savings can be made at little or no cost. Small inexpensive changes may have a substantial effect if considered nationwide – a 2°C lowering of temperature in winter in all buildings in the UK would save about 11 × 10^{16}Joules of energy, i.e. approximately £900 million.

Need to comply with legislation

One of the methods by which governments endeavour to lower their national energy consumption and carbon emissions is through legislation. In the UK this includes the Building Regulations, and in many other countries there are regulations or building codes that have to be complied with. These tend to influence building design, in that they may specify maximum allowable U-values, maximum energy consumption per square metre, the efficiency of building services equipment, and controls. While many of these regulations relate only to new-build, they are increasingly being applied to refurbishments, building extensions and adaptations. In the light of climate change requirements, these regulations will only become stricter in the coming years.

Cost-effectiveness of energy-saving measures

In straitened economic circumstances, clients are often reluctant to invest in anything more than the minimum required to meet compliance criteria, and energy-saving investments can be a 'hard sell'. A company's image may be enhanced if it can promote its 'green' credentials, but it would be naïve to suppose that cost is not always a key driver in decision making. Therefore, when intervention to save energy is considered, many factors must be taken into account: not only the energy savings, but the cost-effectiveness of the energy-saving measures. We need to consider how much energy might be saved, the capital investment required, the payback period, and for how long the measure will be effective. For example, it is easy to reduce the heat losses from a building merely by adding more and more insulation to the walls. Although the addition of insulation saves energy, its cost may be significant, and we have to consider how much energy it really saves over the lifetime of the building, and whether a better investment might be made in more efficient plant or renewable energy sources. A simple illustration of this is given in Chapter 2, while techniques for assessing the cost-effectiveness of measures are examined more closely in Chapter 3.

2 Aspects of building energy use

Environmental requirements in buildings

The energy requirements of a building are largely dependent on the needs of the building occupants and the activities taking place there, and the provision of comfort for the occupants is one of the primary functions of a building. Aspects of comfort include thermal, visual, and acoustic.

- **Thermal** – a state of thermal equilibrium whereby the occupant feels neither too hot nor too cold. The environmental factors involved are air temperature, radiant temperature, relative humidity, and air movement. In well insulated buildings the radiant and air temperatures tend to be very close, although people feel fresher if the air temperature is slightly lower than the radiant temperature.

  A measure of thermal comfort often used is the operative temperature, which in rooms with low air movement (below 0.2m/s) is the average of the air and mean radiant temperatures. Generally people in offices and similar buildings are happy with operative temperatures of 19–25°C, relative humidity 40–70 per cent, and a small degree of air movement, generally below 0.15 m/s. Other factors affecting the comfort temperature include uniformity of temperature distribution, level of activity and clothing. For these reasons, different conditions are required in gymnasia, sports centres and swimming pools. Both domestic and office temperatures have risen gradually over the years as people's style of dress at home and at work has changed. Average temperatures during working hours in offices are commonly 22°C in winter, and sometimes more; air-conditioned spaces often feel too cool in summer. Allowing a greater range of temperatures in summer and winter would lead to large savings in the long term, at no cost. In the UK, the Chartered Institute of Building Services Engineers (CIBSE) guides provide useful information on the appropriate conditions for a range of room and building types.

- **Visual** – occupants should be able to carry out their necessary tasks without visual strain. The quality of lighting is as important as the quantity – light should be of appropriate colour, glare should be eliminated, and lighting should be suitable for viewing computer screens. In addition, lighting for safety (e.g. exits in event of fire, or power cuts) should be provided. Higher lighting levels than needed waste energy, and can lead to eye strain and headaches if endured over long periods.

• **Acoustic** – provision of the appropriate acoustic environment, without intrusion of excessive noise from neighbouring rooms or buildings, or from the building services equipment, e.g. fans, pumps, lifts. The requirements for different building and room types vary depending on the activity to be carried out – school classrooms, for example, require a lower background noise level than engineering workshops.

The requirements for different building types can be found in the appropriate national standards and regulations, or in the guides published by CIBSE and the American Society of Heating, Refrigerating and Air-conditioning Engineers (ASHRAE). Most of the information in this book deals with technical solutions to the problem of saving energy, but even before those solutions are applied there is much that can be done, such as changing people's attitudes and expectations. For example, insulating a house may not lead to energy savings, because the owner may choose to use the same amount of energy as before but enjoy a much higher temperature; similarly, two identical adjacent houses may have energy consumption levels that differ by 100 per cent because of the lifestyle and behaviour of the occupants. There is an ongoing debate concerning thermal comfort, and it is a widely held view that the comfort levels described above are too narrow in range, and do not take account of the ability of people to adapt to their surroundings and the relationship between indoor and outdoor environmental conditions. This adaptive thermal comfort model is gradually being incorporated into standards, and the widening temperature range proposed before heating or cooling plant is brought into operation permits a substantial amount of energy to be saved without incurring further costs.

Where and how energy is used in buildings

Energy is used in a number of ways in buildings, and the most cost-effective interventions naturally result from targeting the sectors where most energy is used. This clearly depends on the building type, the form of construction and the kind of building services provided, but principally on the climate in which the building is located. Human behaviour is also an important factor, and as a first approximation it is often assumed that the impact on energy consumption is based on building/controls/behaviour in the ratio 40:20:40 per cent respectively.

In the UK, most of the energy used in houses is used for space and water heating, while in offices a much greater proportion is used in lighting. In the hot-dry climates dominating much of the Middle East, up to 90 per cent of the energy consumption may be accounted for by air-conditioning, even in housing, whereas air-conditioning is virtually unknown in houses in the UK.

The heating requirement for a building can be divided into two elements: space heating and water heating. Most buildings require hot water for washing purposes in varying amounts depending on the type of building. For offices the requirement is small, as it is only needed for hand washing, while in dwellings substantial amounts of hot water are needed every day for personal, clothing and dish washing.

In climates where space heating is required, it can be further divided into two components: fabric losses and ventilation losses. Fabric losses occur through the roof, walls and floor, and all buildings require ventilation, whether supplied mechanically or naturally. Heat transfer through the building fabric is governed by the thermal transmittance of the elements comprising the outer envelope of the building, known as the U-value. The U-value is defined as the 'average heat-flow rate per area in the steady state divided by the temperature difference between the surroundings on each side of a system' (ISO 7345). Heating the cool outside air to the required indoor temperature incurs a heat loss on the building, known as the ventilation heat loss.

Heat gains to buildings come from a number of sources, external and internal. External gains include solar gains principally through windows,

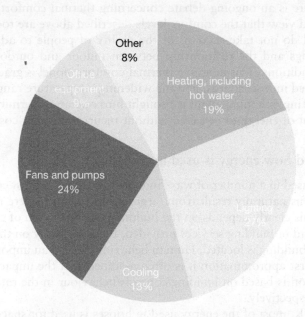

Figure 2.1 Typical energy consumption pattern for a 20-year-old air-conditioned office building in the British climate

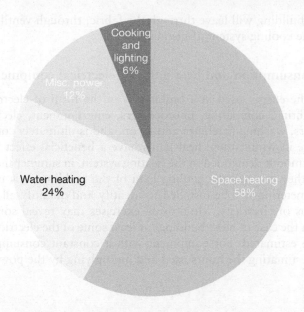

Figure 2.2 Typical domestic energy consumption pattern for a private house in the British climate

heat gains through the building fabric due to higher outdoor temperatures, and through ventilation air. Internal sources include people, electrical equipment such as lighting, computers and other office equipment.

Energy balance

While heat flows through a building are three-dimensional and vary over time, for many buildings and climates we can describe the heat flows sufficiently using a simple heat balance. This assumption is less valid in very hot climates and with buildings containing considerable thermal mass, which has the effect of adding a time lag to the heat transfer process. Some materials, such as stone and concrete, are capable of storing significant amounts of heat, but when considered over a long enough period the residual energy storage is small compared with the flows in and out, and a 'steady-state' energy flow situation may be assumed.

$$\text{Energy in} = \text{Energy out}$$

The time lags induced by additional thermal mass will affect heat-up and cool-down periods in winter and summer, and peak room temperatures; comment is made on this in Chapter 8 when discussing optimum start and stop, but on a simplified analysis, it can be assumed that all the heat that enters the building will leave through the fabric, through ventilation air, or through the cooling system, if installed.

Energy consumption and heat gains of electrical equipment

Much of the energy used in a building is in the form of electricity, which powers lighting, computing, photocopiers, entertainment, electric motors, refrigerators, washing machines and so on, and is ultimately converted into heat. While in winter these heat gains have a beneficial effect in that they offset the amount demanded of the heating system, in summer they are undesirable as they increase the cooling load of the building. It is unlikely that sufficient metering will be installed to identify and quantify all the end-use applications of electricity. Monitoring exercises may reveal some information, but in the case of most buildings at least some of the electrical gains will have to be estimated. For equipment with a constant consumption it is a matter of estimating the hours used and multiplying by the power rating:

$$E_e = W_r \times h$$

Where
E_e is the power consumed
W_r is the power rating
h is the number of hours used.

For monitoring individual appliances, simple hours-run meters can be fitted but their usefulness is limited as they have no facilities for transmitting the data elsewhere.

Where lighting circuits cannot be separately monitored, the installed wattage of a lighting system (without dimmers) can usually be calculated by multiplying the number of lamps or tubes by the individual rating and making an estimate of the hours used.

Example

A large open-plan office has 180 fluorescent tubes of 54 W each. Calculate the annual energy consumption.

Installed wattage = $54 \times 180 = 9.72 \ kW$
Hours of operation (estimated): 5 days a week, 50 weeks a year, 08.00–18.00 hours.
It is further estimated that the lighting is on for 80 per cent of the office hours.
Total hours per year = $10 \times 5 \times 50 \times 0.8 = 2000 \ hours$
Total consumption = $9.72 \times 2000 = 19440 \ kWh$

Reductions in energy consumption can be brought about by using more efficient light sources such as LEDs to produce a lower installed wattage and/or reducing the number of hours of operation at full output.

If counting the light fittings and their individual ratings is not feasible, then an estimate, however approximate, should be made. Heat gains from lighting vary between approximately 8 W/m^2 for efficient systems and 20 W/m^2 for poor ones.

Other items such as computers consume a variable amount of power depending on their state; a laptop may use 25 W in sleep mode, 100 W working in screen-saver mode, and 200 W with the screen on and hard drive in use. For such appliances it is necessary to know both the consumption and the runtime at each power level. In order to minimize energy consumption from appliances, the most effective approach is to specify low-energy appliances in the first instance and use them for the minimum time. Low-energy appliances, including refrigerators, washing machines, etc., can be identified by the 'Energy Star' system (US) and Energy Ratings system used in the UK and the European Union. In the latter scheme a label specifies the consumption and indicates a rating of A–G (A** is now available for refrigerators) in descending order of efficiency.

Internal heat gains also include those from the occupants, ranging from just over 100 W for sedentary activities to over 400 W for heavy manual work. The accuracy of any human heat gains is limited, as it is not always possible to know the exact number of occupants, or their activity level, at any time. Also, when a building such as an office is heated in the morning prior to opening, there are no heat gains from people, so the largest heating demand will tend to occur when there are no 'free' gains to be taken advantage of.

In the absence of any information concerning the rating of computers and other office equipment, then heat gain allowances for office equipment range from 5–15 W/m^2 (see Table A5 in Appendix 2).

Most equipment in a building is not in constant use and to allow for this intermittency, diversity factors are used. A 200 W printer in use for the whole 8 hours of a working day would produce a total heat gain of 1600 Wh. Applying a diversity factor of 0.6 (i.e. it is in use 60 per cent of the time) reduces the heat gain to 960 W.

Heat gains from cooking appliances are more difficult to assess, particularly in large non-domestic kitchens, because exhaust hoods remove a significant proportion of the heat directly. These are latent and convective gains as most of the radiant gains enter the room. Except for restaurants and factory canteens, the heat gains from cooking are likely to be insignificant compared to the other heat flows.

Example of heat-loss calculation

The U-value is a measure of the heat transfer through a building element such as a wall, and is measured in W/m^2K. It can be considered to comprise three main thermal resistances: the inner surface resistance, the resistance to conduction of the wall itself, and the outer surface resistance. The surface resistances can be obtained from the CIBSE and the ASHRAE guides; the wall resistance is the sum of the resistances of the individual layers; the resistance of each layer is simply the thickness t_n divided by its thermal conductivity k_n.

$$U = \frac{1}{R_t}$$

$$R_t = R_{si} + R_w + R_{so}$$

$$R_w = R_1 + R_2 + \dots R_n$$

$$R_n = \frac{t_n}{k_n}$$

For a single-leaf brick wall in a location with 'normal' exposure, i.e. neither highly sheltered nor highly exposed such as in a valley or on a hill top, the following values are used:

Without insulation
    R_{so} = 0.055 m^2K/W
    R_{si} = 0.123 m^2K/W
    t_w = 105 mm, k_w = 0.44
    R_w = 0.105/0.44 = 0.238
    R_t = 0.123 + 0.238 + 0.055
    R_t = 0.417 m^2K/W
    U = 2.4$Wm^{-2}K^{-1}$.

With insulation
Adding 50 mm of insulation with thermal conductivity k = 0.035 gives:
    R_{ins} = 0.05/0.035 = 1.428
    R_t = 0.123 + 0.238 + 1.428 + 0.055 = 1.845
    U = 0.54 $Wm^{-2}K^{-1}$.

Adding a further 50 mm of insulation gives:
    R_{ins} = 0.1/0.035 = 2.856
    R_t = 0.123 + 0.238 + 2.856 + 0.055 = 3.273
    U = 0.305 $Wm^{-2}K^{-1}$.

It is clear from the example that the benefit of adding insulation is subject to a law of diminishing returns. The first 50 mm of insulation reduces the U-value by 77 per cent, while the second 50 mm reduces it by only a further 30 per cent. Figure 2.3 shows the effect on the U-value of adding different thicknesses of insulation to a plain brick wall. Considering the energy savings (in terms of energy cost) over the lifetime of the insulation, and balancing that against the cost of purchasing and fitting the insulation, it becomes evident that beyond a certain point we are investing large amounts of capital for little effect on running costs. While complex formulae have been developed to estimate the economic thickness of insulation in different circumstances, it is difficult to predict how the energy cost will change and therefore their usefulness is limited. A sizeable change in fuel costs has a profound effect on the economics of insulation; as the fuel price increases relative to installation cost, the optimum economic thickness increases. Even for low insulation cost and high energy cost, adding more than 150–250 mm of insulation to an existing element is rarely a worthwhile investment for buildings in the UK and similar climates. For regions with lower heating loads the economic thickness is correspondingly less.

Example of insulation payback calculation

Consider a detached bungalow located in Eastern Scotland, with floor plan 10 m × 7 m, walls 3 m high, with glazing constituting 40 per cent of the wall area, of single-leaf brick construction, with no insulation. Heating is provided through a gas-fired boiler.

The number of degree days at this location is 2,234. The U-value of the walls is 2.4 $Wm^{-2}K^{-1}$. Using the simplified version of the degree-day method (described in Appendix 1) the Annual Energy Consumption (AEC) to compensate for heat losses through the walls only, with a 60 per cent efficient boiler, is given by:

AEC = 2.4 × 61.2 × 2234 × 0.024 × 0.7/0.6 = 9187 kWh, which at 5 p/kWh gives an AEC related to the walls of AEC = £460/year.

Adding 50 mm insulation reduces the U-value to 0.54 $Wm^{-2}K^{-1}$.

AEC = 0.54 × 61.2 × 2234 × 0.024 × 0.7/0.6 = 2067 kWh, giving an AEC of £103/year.

This results in savings of £357/year. Adding 50 mm insulation costs £8/m² (£490) and has a payback period of 1.37 years. Alternatively, if 100 mm insulation had been added from the outset instead of 50 mm the energy consumption would be:

AEC = 0.305 × 61.2 × 2234 × 0.024 × 0.7/0.6 = 1167 kWh, giving
AEC = £58/year and savings of £402 over the original wall. The cost
of the insulation and installation is approximately double (£980),
giving a longer payback period of 2.4 years and overall savings
dependent on the useful lifetime of the measure: insulation will generally last 20 years or more with the same effectiveness.

In this instance the thicker insulation gives a longer payback period
and may therefore appear less attractive; however, payback period is
only one of the criteria that may be used to make a decision (see
Chapter 6).

The effect of insulation on energy costs and payback depends on the
nature of the initial wall construction. For a brick cavity wall without
insulation, the figures used in the previous example become:

original, no insulation, U = 1.38 $Wm^{-2}K^{-1}$, energy cost £264
total 50 mm insulation, U = 0.31 $Wm^{-2}K^{-1}$, energy cost £59,
    payback period 2.4 years
total 100 mm insulation, U = 0.22 $Wm^{-2}K^{-1}$, energy cost £42,
    payback period 4.4 years.

The better insulated the wall already is, the longer the payback period
for the insulation will be. In the examples given the payback periods
are relatively short, and the measure would be expected to last for 20
years. Therefore the total net savings (i.e. lifetime savings minus cost
of insulation) in each case over 20 years are as follows:

single brick, 50 mm £6,650
single brick, 100 mm £7,060
cavity brick, 50 mm £3,610
cavity brick, 100 mm £3,460.

Once the specific heat-loss coefficient and annual heating energy consumption have been determined, steps to identify appropriate energy-conserving
measures can be taken. Firstly, the routes of greatest heat loss should be
identified. If information on ventilation rates and the U-values of the
elements of the outer envelope of the building are available, the heat loss
may be broken down into its individual components of loss through the
building fabric, ventilation, infiltration, and losses through boiler ineffi-
ciency. These heat losses can be further broken down into walls, windows,
roof and ground floor. Later chapters will show how some of these measurements or estimates can be made, but for the present we will assume the
data below.

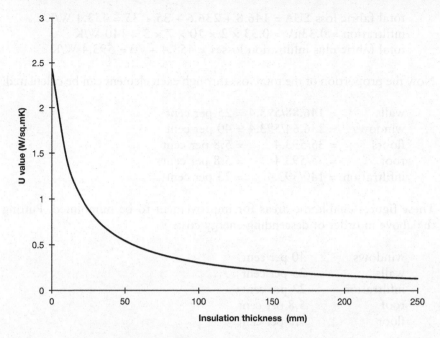

Figure 2.3 Effect of insulation thickness on U-value

Continuing with the example of the bungalow used earlier, the U-values of the outer envelope of the building are as follows:

walls	2.4 W/m²K
windows	5.8 W/m²K
ground floor	0.5 W/m²K
roof	0.5 W/m²K.

The total heat loss is made up of the losses through all these routes and through infiltration, which is here estimated at two air changes per hour.

Boiler efficiency is 60 per cent.

The contribution of each element to the total heat loss and energy bill can be calculated as below:

wall area = (10+10+7+7) × 3 × 0.6 = 61.2 m²
window area = (10 + 10 + 7 + 7) × 3 × 0.4 = 40.8 m²
specific losses from walls, ΣUA = 2.4 × 61.2 = 146.8 W/K
specific losses from windows = 5.8 × 40.8 = 236.6 W/K
specific losses from floors = 0.5 × 70 = 35 W/K
specific losses from roof = 0.5 × 70 = 35 W/K

total fabric loss ΣUA = 146.8 + 236.6 + 35 + 35 = 453.4 W/K
infiltration = 0.33nV = 0.33 × 2 × 10 × 7 × 3 = 140 W/K
total fabric plus infiltration losses = 453.4 + 70 = 593.4 W/K

Now the proportion of the total loss through each element can be calculated:

walls	= 146.88/593.4	= 25 per cent
windows	= 236.64/593.4	= 40 per cent
floors	= 35/593.4	= 5.8 per cent
roof	= 35/593.4	= 5.8 per cent
infiltration	= 140/593.4	= 23 per cent

These figures enable the areas for improvement to be pinpointed. Putting the above in order of descending energy cost:

windows	40 per cent
walls	25 per cent
infiltration	23 per cent
roof	5.8 per cent
floor	5.8 per cent

The elements that need to be targeted for improvement are those showing the greatest heat loss: the windows, walls and infiltration. While the U-value of the roof is not particularly good and may be relatively simple to improve, since only 5.8 per cent of the heat loss occurs through that route, it is unlikely to be cost-effective. For relatively simple buildings a spreadsheet model can be produced to carry out calculations of this kind. The heat loss through a wall can be reduced progressively by adding more and more insulation but there is a limit to the useful thickness of insulation that may be applied, which depends on a number of factors, including cost, fuel price, climate, the building construction and its use. Figure 2.4 shows a number of scenarios: low energy cost, high energy cost, low insulation cost and high insulation cost. Even for low insulation cost and high energy cost, beyond 150–250 mm of insulation, the investment is not worthwhile in financial terms, while for low energy cost and high insulation cost in this example, the optimum insulation thickness is about 30 mm. The difficulty for the energy manager is that although present energy and insulation costs may be known, it is difficult to predict how the energy cost will change, but we can be fairly certain that it will continue to increase in the long run. A change in fuel costs has a significant effect on the economics of insulation, as Figures 2.4 and 2.5 show.

Although some of the energy-consuming (and heat-producing) activities in the building may be assessed from power ratings, it may be necessary or useful to make measurements of the conditions and actual energy use; meth-

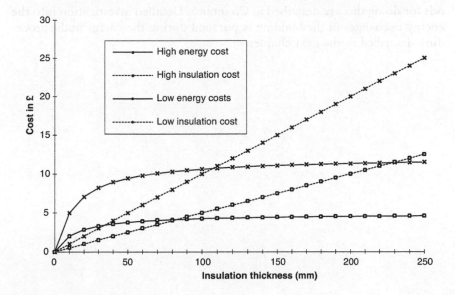

Figure 2.4 Sample scenarios

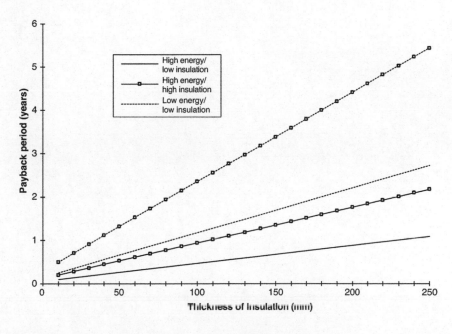

Figure 2.5 Payback period in relation to insulation thickness

ods for doing this are described in Chapter 5. Detailed investigation into the energy exchanges in the building is pursued during the energy audit procedure described in the next chapter.

3 Energy audits

The energy audit

An energy audit is an essential tool of energy management. It is an investigation into the energy use in a building and involves a detailed analysis of the energy flows through the building. It is carried out in order to pinpoint those areas where improvements in the building fabric, services, controls and management can be made, thus making it possible to identify actions that will lead to savings in energy and costs. Other aims include:

- reduction in carbon emissions
- improved environmental conditions for the occupants of the building
- the development of a system for recording energy use
- the development of monitoring and targeting schemes.

The energy audit is not only a useful tool but, under the Energy Savings Opportunity Scheme (ESOS) programme in the UK, mandatory four-yearly energy audits will be required for organisations above a certain size (see Chapter 7).

Heat energy enters or leaves a building as a result of a temperature difference between the inside of the building and the external environment, by transmission through the building fabric or as a result of air movement into or out of the building.

Additional energy enters the building from a number of sources: fuels such as oil, coal, gas, electricity; and 'free' sources such as solar radiation and people. Heat may leave the building via a number of routes:

- transmission through the fabric
- infiltration and ventilation losses
- flue losses
- cooling tower losses.

The energy is used to power:
- heating equipment
- lighting

- cooling equipment
- fans
- pumps
- cooking appliances
- refrigerators
- small appliances (e.g. computers, copiers, fax machines)
- home entertainment
- larger items of machinery
- lifts and escalators.

The energy audit process may be divided into a number of phases:
- pre-survey information and data collection
- the building survey
- analysis of the data collected
- formulation of energy-saving solutions
- reporting of results.

Pre-survey data collection

It is important to collect as much information as possible about the building before embarking on a walk-through survey. A number of sources of information may be available. The utility bills for the building will yield useful information on the amount of energy purchased and the tariffs paid; sub-metering will facilitate a more detailed analysis of where energy is used. If sub-metering is not available, then this should be noted as a potential area for improvement. Even if a meter is not used for billing, regular readings, taken manually if necessary, should be incorporated as part of the monitoring strategy. In addition to showing the number of units consumed, electricity bills will also show maximum demand charges (see Chapter 8), and further analysis of these and of the operation of the building should reveal whether load shifting is feasible as a means of reducing these charges. A series of bills going back a number of years will also enable long-term trends in energy usage to be revealed. Many buildings now have Building Management Systems (BMS) which record detailed energy-use patterns. Half-hour metering will be in use in many cases and will provide much finer detail on daily energy profiles.

Any plans, elevations or technical data on the building should be obtained. These will provide useful information on dimensions, construction materials, (possibly U-values), the building services plant, including type and size, and may also include details of the control strategy.

The information collected during this first phase may include:

- the utility bills – gas, electricity, oil, solid fuel, water
- plans and elevations of the building, including building services systems
- climate and location of the building

- information on the controls, zoning and BMS
- information on the structure of the building and U-values
- information on the building's purpose, hours of work and operation.

The building survey

The pre-survey data collection will give a general idea of the size and layout of the building and the type of plant used. However, a walk-through survey of the building is essential, and will provide much additional information which cannot be gleaned from the plans. One purpose of the walk-through is to determine whether the building is as stated on the plans; often changes are made such as extensions, new ventilation plant, additional insulation, and the like, which may not necessarily be reflected in the available information. The condition of the building should be noted too: fabric defects such as badly fitting windows and doors contribute to energy waste, as does missing or damaged lagging on pipes, or faulty insulation on roofs and walls. The standard of cleanliness of light fittings should also be noted – if lighting output is related to daylight level, dirty fittings may result in 20 per cent more energy being required. As far as possible, the U-values of the building fabric should be checked to see if they are as stated. This may be difficult or indeed impossible to ascertain without causing damage but interviews with building managers, engineers, caretakers and others can often help fill in missing information and provide useful anecdotal evidence to help build up a complete picture of the nature and operation of the building.

Detailed inspection of plant rooms is essential to confirm the type of plant, ratings and nominal efficiency as well as providing a visual indication of the standard of maintenance. Examples of poor maintenance such as broken valves, leaks, missing lagging and other defects should also be noted here as they have an impact on plant efficiency. The location of meters and sub-meters should be identified. Environmental conditions in the building may be measured during the walk-through by the use of hand-held devices which enable spot readings of temperature, humidity, lighting level and CO_2 level to be taken. While these will only give an indication of the conditions prevailing at the time of the survey, they can be useful in identifying potential problems, for example with temperature control. Other control issues may also be identified during the visit, such as:

- Do the users have manual override for heating controls?
- Are there open windows immediately above working radiators?
- Can the users adjust light levels?
- Where are the light switches?
- Are computers and other items such as photocopiers switched off at night or at weekends?
- Is there manual control of ventilation?
- Are fans switched on at night for night cooling?

There may be opportunities (with the employer's permission) to interview the building users or issue questionnaires. In the absence of extensive sub-metering, information to help with the assessment of individual items of energy consumption should be collected, noting what items of plant are running, whether lights are on, and what the policy is on such matters. Even with sub-metering the data obtained may only go down to ring-circuit level, not to the level of individual appliances, therefore these observations are important in identifying what items of plant are in use.

The hours during which the building is operated should also be noted, as should the company's policy regarding late-night and weekend working – if some employees have to work late, for example, does the whole building have to be heated, or is it appropriately zoned?

Analysis of the data collected

The form and extent of the analysis carried out depends on a number of factors: the depth of the audit required, the nature of the building, the degree to which the energy consumption data can be disaggregated into end-uses, and the needs of the client.

It may be that sufficient information is obtained from the pre-survey and the walk-through to make a detailed energy analysis of the building, especially if BMS data is available. Alternatively, additional data may be acquired through a monitoring exercise. The equipment required for this is described in Chapter 5. For large complex buildings it may also prove worthwhile to model the performance using proprietary software. A number of packages of varying complexity are available, allowing simple steady-state heat-loss calculations or detailed hourly simulations of heating, ventilation and air-conditioning (HVAC) performance to be made. Provided reasonably accurate input data is available and the model is suitably validated, it is possible to simulate the effect on energy consumption of changes to the building such as upgrading the fabric insulation and improving plant efficiencies. Where possible, the measured or estimated energy consumption should be compared with that of other buildings having the same function. In the UK, the Carbon Trust issues a large number of publications outlining typical and best-practice energy-use patterns in a large range of building types, including offices (e.g. ECON 19), academic buildings, theatres, health care buildings and many others. Comparisons are normally based on kWh/year/m^2.

Formulation of energy-saving solutions

The analysis of energy consumption in the building will identify where energy use is high and where there is waste. The means by which waste can be reduced are various, and range from no-cost solutions, such as changes to occupant behaviour; through low-tech solutions, such as adding blinds

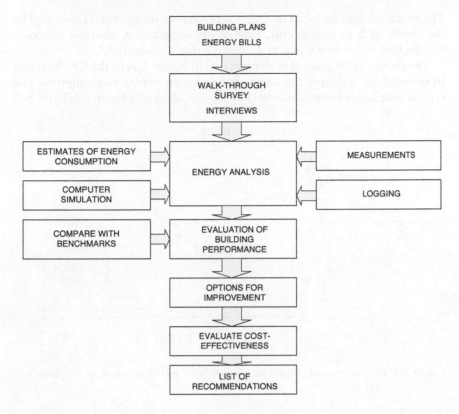

Figure 3.1 Summary of the energy audit process

to the windows, to highly engineered and costly measures, such as the installation of a combined heat and power plant. The selection of measures to be adopted will depend on technical considerations, their cost-effectiveness, capital cost, and the overall energy management strategy of the organisation; examples of such measures are discussed in Chapter 4.

Reporting

A report should be made to the client outlining:

- the present state of the building
- an analysis of current energy use
- identification of areas of waste and where energy can be saved
- details of the kinds of intervention which will reduce energy use
- details of the savings possible
- the cost-effectiveness of the methods recommended.

The measures may be listed in order of preference using criteria specified by the client, such as lowest cost, ease of implementation, shortest payback, Net Present Value (see Chapter 6), lowest risk, as required.

The energy audit process is summarized in Figure 3.1. In the UK, heat loss from buildings accounts for some of the greatest energy consumption, and typical heat losses from houses and offices are shown in Figures 3.2 and 3.3.

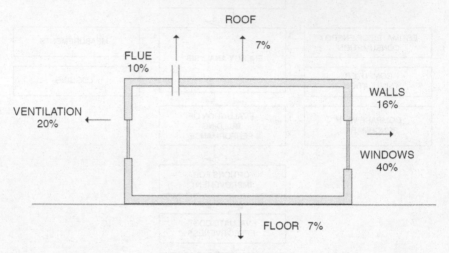

Figure 3.2 Distribution of annual heat losses from a typical house in the British climate

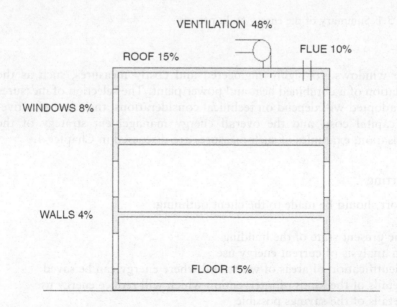

Figure 3.3 Typical heat losses from an office building in the British climate

Analysis of the energy consumption into different end-uses is essential, although the level of detail achievable will vary depending on the input data available. It is then useful to have some sort of yardstick against which to assess the performance, preferably buildings of a similar type in the same climate. For the UK, the Carbon Trust has many publications based on surveys and case studies which can be used for such comparisons; also useful is the Chartered Institution of Building Services Engineers (CIBSE) Guide part F, Energy Efficiency. For other countries, the amount of information available varies considerably. These comparisons are important as they give an indication of the scope for improvement in the different areas of energy use such as heating, cooling, lighting etc., an example being given in Case study 1. A range of measures which might be applied to improve the energy performance is investigated in Chapter 4, where some calculations are presented; the economic assessment methods are discussed in Chapter 6. A list of the best options can then be drawn up and presented to management in the form of an energy audit report. The final recommendations for improving the building's energy performance are based on technical and economic appraisal of the potential intervention measures.

The energy audit report should include:

1   A description of the building or buildings – climate, location, orientation dimensions, materials, purpose (office, workshops, labs, etc.), hours used.
2   A description of the heating/cooling/lighting systems, air-handling units, boilers, fittings, controls.
3   Thermal comfort – state whether conditions are acceptable and present results of any measurements made.
4   Energy consumption – bills, estimates. If not available, state what measurements or monitoring exercises may be needed.
5   Comments on specific points about the operation of the buildings, e.g. does everything work as it should, is it well maintained, etc.
6   Comparisons with buildings of a similar type and use.
7   A list of where energy is being wasted and appropriate measures to enable savings to be made.
8   A series of calculations based on 6 to show the cost-effectiveness (or otherwise) of measures.
9   Specific recommendations based on 7 and 8, listed according to the criteria specified by the client.

Check list for energy audits

Pre-inspection

•   location
•   external environment
•   building form and orientation

- use of building
- floor plans and elevations
- drawings of services layout
- areas, floor-ceiling heights
- maintenance records
- alterations and improvements
- energy sources and tariffs
- determine whether the building is owner-occupied
- determine whether the owner or occupier pays the fuel bills
- determine whether heating/electricity are flat-rate or metered.

Fabric survey

- roof – type, condition, insulation, condensation roof lights
- walls – type solid, cavity material, thickness, insulation, condensation
- floors – type, insulation
- doors – draught stripping, opening type, door closers, revolving doors
- windows – frames, single/double condition, how well fitting, draughty, openable area and orientation, curtains, blinds coatings
- conservatories/atria – orientation type, size, heating or not, what used for, blinds, ventilation.

Energy supply

- electricity
- metered, sub-metered, on site generation chp
- gas – piped, storage tank
- liquid/solid fuel – storage.

Space and water heating

- energy source – boiler plant type, age, condition efficiency heat recovery location and type of emitters, zoning, controls, standard of maintenance
- assess opportunities for using renewable energy sources
- check thermal comfort
- hot water – energy source, type, local or central, condition, temperatures, control.

Ventilation

- Natural or mechanical ventilation, fans, size, filtration, condition, ductwork, control, maintenance, heat recovery, recirculation.

Cooling

- Natural or mechanical, type and size of plant, control, heat recovery.

Lighting

- quality of daylighting
- type of lamp, installed wattage, age, condition, control, cleanliness.

Small equipment

- Computers and other office equipment – type, efficiency, energy star-ratings, hours of use.

Analysis of impact of potential changes – building

- assess results of upgrading fabric – walls roofs floors windows
- assess cost and energy savings, and other benefits such as improved comfort
- windows – assess condition of window frames and need for upgrading; this will factor into the case for double glazing
- assess opportunities for changing window size
- assess effect of blinds or shading on cooling load
- doors – assess opportunities for reducing heat losses – revolving doors, air curtains, door closers.

Analysis of impact of potential changes – energy and HVAC

- energy – assess effects of switching fuels, suppliers
- assess effects of changing boilers and/or heating system
- consider use of heat pumps/CHP
- hot water – assess effect of changing to/from point of use heaters
- ventilation – assess potential for heat recovery, changing to natural ventilation
- cooling – assess potential for improving efficiency, assess potential for natural cooling
- lighting – assess potential for improving efficiency with new luminaires or improved controls.

Refurbishment considerations

Although an energy audit may reveal several measures that need to be taken to improve a building, opportunities for instigating energy management procedures do not present themselves continuously, i.e. it may only be cost-effective to install insulation, upgrade plant, lighting, etc. when other building works are planned, because disruption to the normal working of the building would result in loss of output. In addition, company policy may dictate that plant may only be replaced when it has come to the end of its useful life, and fabric enhancement may only be considered viable during a major refurbishment. A refurbishment programme offers some of the best opportunities to upgrade the building fabric and install more energy-efficient HVAC and lighting plant, and may be carried out for a number of reasons:

- change of use of building
- moving into an old vacant building
- plant has reached the end of its useful life
- building fabric is in a rundown condition
- conditions in building no longer acceptable

The level of refurbishment may include: replacing carpets, partitions, light fittings, boilers, heat emitters, the entire heating system, ventilation system; removal and replacement of finishes and insulation on outer walls; fitting false ceilings, raised floors, new internal walls; adding or removing air-conditioning units; replacing control systems and fitting a BMS. The lifetime of different elements of the building – structure, services, etc., varies (Table 3.1) and dictates the extent of the changes that are considered feasible.

Table 3.1 Lifetimes of different elements of a building

Element	Lifetime (years)
Structure	50+
Building services	20
Cabling	12
Office equipment	6–8
IT hardware	2–4

Example of analysis of heating data

As shown in Figures 3.2 and 3.3, heat losses from the building fabric show marked variations depending on the type of building and construction method. However, if the heating in a building in winter is correctly controlled, there should be a strong correlation between energy use and external temperature, therefore a plot of heating energy use against temperature should yield a straight line. Such a plot on its own only shows that consumption increases in winter and reveals little in the way of detailed information. A more meaningful examination of the data can be made using the concept of degree days described in Appendix 1. An example of this is given below. It is based on a building in the UK climate that needs to be heated for about half the year. In this instance mains gas is used for both heating and cooking, but the figures are not disaggregated; monthly bills are available, as is monthly heating degree-day data for the location. Utility bills tend to be issued on a quarterly or monthly basis, but if a BMS is installed, or if manual readings are taken, then weekly or even daily data may be available for a much finer level of detail. Many large users of energy in the UK will now have automated half-hourly metering as standard.

Table 3.2 shows the monthly gas consumption for the building. The gas consumption in cubic metres is divided by its calorific value (39.1MJ/m³) and by 3.6 to convert to kWh. Table 3.3 and Figure 3.4

Table 3.2 Monthly gas consumption

Month	Total gas consumption (m³)
Jan	932
Feb	824
Mar	642
Apr	461
May	83.6
Jun	91.7
Jul	80.3
Aug	74.1
Sep	86.7
Oct	558
Nov	734
Dec	786
Annual total	5853.4

Table 3.3 Monthly energy consumption and degree days

Month	Total gas consumption (kWh)	Degree days
Jan	10,123	576
Feb	8,963	453
Mar	6,982	324
Apr	5,015	272
May	909	0
Jun	997	0
Jul	873	0
Aug	806	0
Sep	943	0
Oct	6,063	327
Nov	7,987	434
Dec	8,546	524
Annual total	49,661	2,910

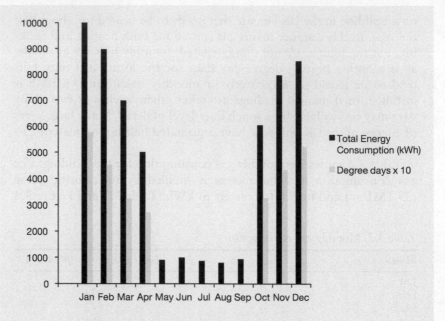

Figure 3.4 Energy consumption and degree days (× 10)

show the kWh consumption along with the degree days at the site for the same period. If a BMS is used, external air temperatures may be recorded and accurate values of the local degree days can be calculated; alternatively, degree-day values may be obtained from a number of sources as described in Appendix 1. Space heating and cooking both use gas but are not separately metered, therefore the values in Figure 3.4 include both heating and cooking energy. Water heating is carried out separately, by electricity.

The end-use consumption is disaggregated as follows. During the months May to September inclusive, the heating is switched off entirely and no heating can be used in the building, even if the conditions are cool enough to require it. Therefore any gas consumed during these months is used only for cooking. As may be seen from Figure 3.4 and Table 3.2, the amount of energy used for cooking does not change significantly from month to month; taking the average gas consumption during the summer months will give a reasonable estimate of the monthly cooking energy consumption throughout the year. Subtracting this from the total gives the monthly gas consumption due to heating (Table 3.4 and Figure 3.5).

The revised gas consumption (for heating only) can now be plotted against the degree days (Figure 3.6).

Table 3.4 Monthly heating energy consumption and degree days

Month	Heating energy consumption (kWh)	Degree days
Jan	9,217	576
Feb	8,057	453
Mar	6,076	324
Apr	4,109	272
May	0	0
Jun	0	0
Jul	0	0
Aug	0	0
Sep	0	0
Oct	5,157	327
Nov	7,081	434
Dec	7,640	524
Annual total	*47,339*	*2,910*

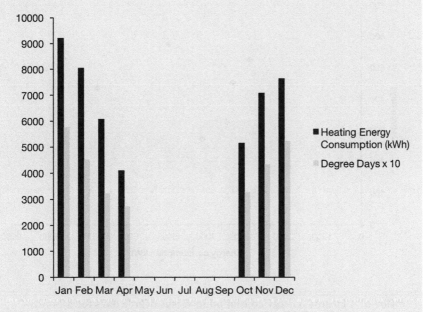

Figure 3.5 Heating energy consumption and degree days (× 10)

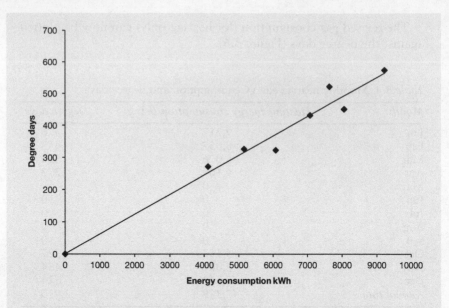

Figure 3.6 Heating energy consumption versus degree days

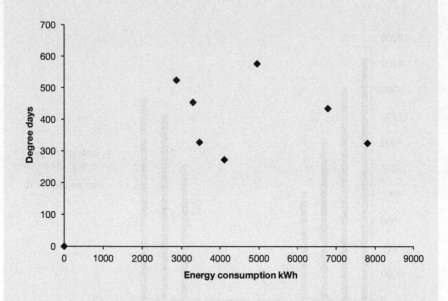

Figure 3.7 Heating energy consumption versus degree days for a poorly-
controlled building

The regression routines as found on standard spreadsheet packages can be used to obtain the best straight line (Figure 3.6). The slope of this graph gives energy use per degree day. Dividing this by 24 gives kW/K, the specific heat loss or total heat-loss coefficient (TLC) of the building. In practice there will always be a degree of scatter on the graph, associated with the normal behaviour of people in buildings, such as the periodic opening of windows or doors, and variations in the number of occupants and the use of lights, machinery or electrical equipment. A number of reasons associated with the design and operation of the building and the building services also serve to weaken the link between consumption and conditions. Examples of these are: extreme time lags and poor control of the heating system, an oversized or undersized boiler, and poor control of the boiler air supply. The scatter of points may be so great that that it becomes difficult or even meaningless to attempt to find the best straight line. A high degree of scatter may indicate a serious problem either with the heating system itself or with its controls, such as a broken sensor or actuator (Figure 3.7). These matters should be investigated thoroughly before instigating other energy-saving measures. Problems such as these will also tend to manifest themselves as poor temperature control experienced by the occupants. Since the main function of the building services is to keep the occupants comfortable, if it fails to achieve that, any energy used is being wasted. The regression line can be drawn on the graph by using one of the standard spreadsheet packages or software within the BMS. In a general sense, the points above the line suggest that energy was wasted for some reason, while those below the line imply good practice. Those occasions producing points below the line should be investigated further so that the conditions producing below-average consumption can be repeated if possible, and those above the line should be studied so that the reasons for excessive energy usage can be identified. It is also possible to use a redrawing of the graph, using only those points on the line or below it to produce new targets (see cumulative sum – CUSUM – analysis Chapter 6).

4 Techniques for reducing energy consumption

The general strategy for reducing energy consumption should be:

- repair faults and ensure good maintenance of plant
- reduce energy loads where possible
- use efficient plant to service the loads
- use efficient sources of energy to operate the plant, which may include renewables.

Once the sources of energy waste have been identified through the energy audit, techniques for reducing this waste must be found. These can conveniently be placed under the following headings:

- building fabric
- building services plant
- controls
- management of the building
- energy supply.

Improvements to the building fabric and building services plant are considered in detail in this chapter, and some calculations showing the energy savings possible are presented in the case studies at the end of the book. Controls are covered separately in Chapter 8. While the availability of finance plays a considerable role in determining the kinds of intervention possible, practical considerations also dictate to a large extent the kind of changes that can be made. Many commercial buildings are not owned by the occupiers and this severely limits the interventions that are possible. Furthermore, some of the measures may only be suitable for individual detached buildings and there may be both practical and management difficulties in applying them to terrace houses or blocks of flats. Flats in particular present problems in upgrading them for energy efficiency, as they may have centrally supplied heating or restrictions on the changes that can be made to the building fabric.

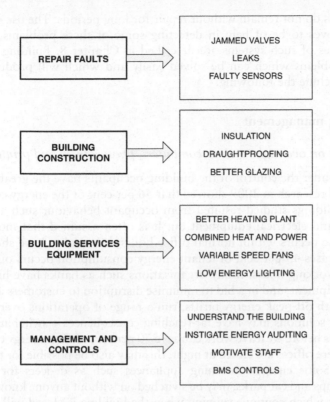

Figure 4.1 Typical techniques to help reduce energy consumption

Repairing faults

Any approach to improving building performance must first ensure that the existing fabric is sound and that the heating, cooling, ventilation and lighting plant, and their associated controls, are operating correctly. Any damage to the building fabric such as broken windows, loose tiles, and damp or missing insulation, will compromise the energy performance and should be attended to as part of the normal building maintenance routine. Other typical problems include dripping taps, leaking pipes, jammed or broken control valves, broken or inaccurate sensors, and clogged filters. Good maintenance is essential to keep up a high level of energy efficiency, and also to maintain thermal comfort and air quality. Dirty ductwork will restrict flow and allow unclean air to enter the workplace, having an adverse effect on the occupants of a building. Correct burner settings and an adequate supply of combustion air are also essential in maintaining safe and efficient operation of boilers, as is the regular checking of flues. A systematic maintenance and repair regime should be instigated to ensure that faults do not remain without repair for long periods. The use of a BMS often proves to be of help in detecting some of these problems, and the advantages of such systems are described in Chapter 8. Building management problems which can be solved easily and which will produce quick results include the following.

Building management

Items left on overnight such as computers, photocopiers and printers

Factors under the control of the building occupants have the greatest influence, and research in 2009 showed that 56 per cent of the energy consumed in six buildings studied resulted from occupant behaviour such as leaving lighting and electrical equipment on. It is often assumed that most of the energy use occurs within normal office hours, but studies have shown that in some cases 45 per cent of tenant energy consumption occurs outside the normal working hours. Many organisations such as banks have high levels of IT equipment, and in a bid to minimise disruption to customers and coordinate with off-peak energy tariffs, run a range of operations overnight.

Simple solutions may have far-reaching consequences – switching off our computers before we leave work can save 70 per cent of their energy consumption. Where offices are closed at night, this may also be possible for telephone systems. Some energy-consuming appliances such as de-icers for external steps, ramps and car parks, may be switched on without anyone knowing, and escalators which continue running when the building is closed will consume considerable amounts of energy over a year. The influence of human behaviour on energy consumption is significant, and some studies suggest that it can have a greater impact than U-values. There are many reasons for poor human behaviour in relation to energy use, such as disengagement, lack of interest, and a lack of a connection between the user and the energy bills. Methods for improving motivation and engagement are explored in Chapter 6.

Building fabric

Heating, cooling and lighting constitute the major energy uses in most buildings, and are largely determined by the climate, but also by the basic form and construction of the building, which normally cannot be altered. However, major refurbishment may present opportunities to add a conservatory or atrium, add roof lights or light shelves to improve daylighting, or even to change the proportion of glazing on the façade. All of these could enhance the building's energy performance, but where they also increase the internal space available, there is a danger that the overall energy consumption will increase. The greatest opportunities for fabric changes lie in adding thermal insulation, replacing single glazing with double glazing, and reducing infiltration by installing draught-proofing materials.

Insulation

In many instances, adding thermal insulation to the inside of a wall is the easiest way to improve the building fabric but it reduces the usable floor space to a small extent, and can prove problematic around details such as windows, or where there are radiators against the walls. Cavity insulation or externally applied insulation panels do not have this drawback, but external insulation is expensive as it is required to withstand the rigours of the outdoor environment. Externally applied thermally insulated cladding is particularly useful for high-rise flats; it is easier to install than internal insulation to each flat, and is cost-effective. It also retains the thermal mass of the building, which helps to mitigate temperature variations; the effect of this thermal mass is lost when internal insulation is used. In houses, the easiest location to add insulation is in the loft space, as the process of installation causes little disruption, and a thick layer may be used without intruding on the living space. Where there are rooms in the attic, the insulation must follow the roof line, which increases the cost of installation. Commercial buildings are more likely to have flat roofs, which can be more difficult to insulate. Insulation of ground floors in existing buildings is easiest where there are suspended floors, provided there is sufficient crawlspace underneath; there are a variety of arrangements of insulation, utilizing rigid insulation boards, or loose-fill mineral wool supported by netting (Figure 4.2). Insulation of solid floors is more difficult, but strong rigid insulation boards can be laid on top of concrete floors, and wooden flooring laid over

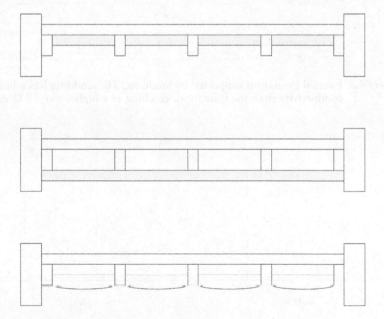

Figure 4.2 Insulating below suspended floors

the top. This of course raises the floor level, leading to significant additional work, disruption to the occupants, and costs. It is sometimes found that insulation fails to deliver the anticipated energy savings, for a number of reasons: the occupiers may allow internal temperatures to rise, since they can now afford more comfortable conditions; the effect of supporting members on the U-value may not have been allowed for; after some time loose-fill insulation may settle, leaving un-insulated areas. The effect of studding to support the insulation is sometimes ignored (Figure 4.3) – it has a higher thermal conductivity and increases the effective U-value, however if the correct U-value calculation is used, this will be taken into account. Cavity insulation may fill all or part of the cavity (Figure 4.4) and produces different

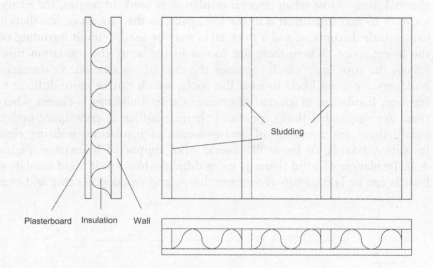

Plasterboard   Insulation   Wall   Studding

Figure 4.3 Internal insulation supported by studding. The studding has a higher
conductivity than the insulation, resulting in a higher overall U-value

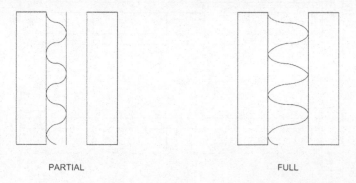

PARTIAL   FULL

Figure 4.4 Partial and full cavity wall insulation

U-values. A 100 mm cavity with partial fill of 50 mm insulation has a thermal resistance value of 1.38 m^2K/W, while filling the cavity completely with insulation would produce a value of 2.5 m^2K/W. Normally in a retrofit situation, beads are blown into the cavity to produce full cavity insulation.

Thermal bridging

In all building elements there may be thermal bridges: areas of low thermal resistance connecting the inner and outer surfaces, through which heat passes relatively easily. These can significantly lower the effectiveness of insulation in comparison with the estimated reductions in energy consumption. Such bridges typically occur at corners, window sills and soffits. Careful detail design is needed in order to avoid these (Figures 4.5 and 4.6), but in many instances the basic building structure means that they are unavoidable. The thermal bridge through the window frame can be avoided by using frames having a thermal break, although the sill and soffit may still constitute a thermal bridge. One effect of thermal bridging is to produce areas on the outer wall such as corners that are at a lower temperature than the centre of the wall, and these lower temperatures can give rise to mould growth and damage to the building fabric and decorations. A combination of carefully detailed insulation, higher temperatures and adequate ventilation is required to avoid this problem.

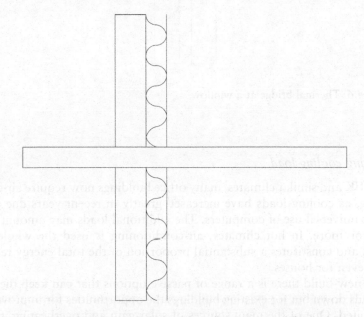

Figure 4.5 Thermal bridge at a balcony. Heat flowing out of the building can bypass the wall insulation unless the floors and ceilings are insulated

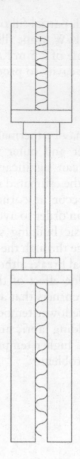

Figure 4.6 Thermal bridge at a window

Reducing cooling load

In the UK and similar climates, many office buildings now require air-conditioning, as cooling loads have increased greatly in recent years due to the almost universal use of computers. The additional loads may amount to 10 W/m² or more. In hot climates, air-conditioning is used the whole year round, and constitutes a substantial proportion of the total energy requirement, even for houses.

For new-build there is a range of passive options that can keep the cooling loads down but for existing buildings the opportunities for improvement are limited. One of the main sources of solar gain and overheating is solar radiation entering through windows. The means of controlling solar gain include the use of internal and external solar shading devices.

Reducing the solar gains by the use of internal blinds and shades is reasonably cost-effective and may help to reduce the peak gain in a south-facing room by 15–20 W/m^2. At a higher level of investment, external shading, fixed or movable, can offer even more improvement. It should not be forgotten that insulation will help to keep heat out as well as reduce heat losses, and at low latitudes roof insulation can help to lower cooling loads considerably, since the bulk of the solar gain is on the horizontal plane.

Energy consumption for cooling can be reduced by using natural or passive methods. The extent to which these can be applied in existing buildings is limited by the form and construction of the building and the overall heat gains to the building. Passive methods can cope with a limited range of heat gains, but if they are not able to eliminate it the cooling load altogether they may be able to reduce it.

Passive or 'natural' cooling in buildings can be provided by:

- Natural ventilation, which removes warm internal air and replaces it with cooler external air. This can be used when external temperatures are lower than those inside, and achieved through open windows and doors. The amount of ventilation and cooling is limited by the temperature difference, wind speed and the size and configuration of openings. As a retrofit measure, opening windows may replace sealed windows and therefore allow some wind-driven ventilation. In moderate climates such as those of the UK it can be effective but in warmer climates more sophisticated methods may be called for as cooling will only occur when the outdoor air temperature is lower than that indoors.
- Thermal mass, which can be used to even out variations in internal and external conditions, absorbing heat as temperatures rise and releasing it as they fall. This is clearly difficult to achieve in a retrofit but internal walls that are covered in insulation or other material can be stripped bare to expose the thermal mass material of bricks or stone. This style is currently fashionable in restaurants in and retail establishments in the UK and enables some reduction in peak cooling load to be achieved.
- Evaporative cooling, such as moisture evaporating from the surface of a building, or the inclusion of water features such as ponds. Again, not an easy measure to implement during a refit, but sometimes fountains can be introduced into atrium spaces to contribute towards cooling. Typically, evaporative cooling is best suited to hot, dry climates as it results in humidification of the air, which would be unwelcome and ineffective in a humid climate.
- Shading devices over windows can reduce the solar gains entering a building by up to 40 per cent in hot climates.

- Reflective surfaces, insulation, green roofs and other measures, while they may not in themselves provide all the cooling required, can reduce the heat gains and contribute something to reducing the cooling demand from any air conditioning systems.
- Passive systems may be inappropriate in some circumstances, such as where high levels of noise or pollutants are prevalent, and where the use of natural ventilation may have to be minimised. In such cases mechanical ventilation may be unavoidable.

Active cooling

Active cooling usually requires some energy input from fans or pumps and can be provided by:

- Earth-to-air heat exchangers (ground coupling), which draw air for ventilation through buried ducts, tubes or a labyrinth using fans. As the temperature 3 m below ground level is practically constant (at around 11°C in the UK), it can be used to substantially reduce ambient air temperature fluctuations, the incoming air being heated in the winter and cooled in the summer. Again, the potential for fitting this into an existing building is limited.
- Open or closed loop water-to-air heat exchangers, which exploit the relatively stable temperature of the Earth to provide water that can cool in the summer and heat in the winter. This method has been used successfully to provide 'free' cool water for the air conditioning system in the Scottish Parliament building in Edinburgh.
- Night-time purging of a building with cool night air, expelling the warm air and cooling the building mass for the following day. This can at least reduce the cooling load if not eliminate it and can, in certain circumstances, be used in existing buildings with few modifications. This is sometimes considered to be a passive system, but where fans are required it may be considered to be 'active'.
- Chilled beam, chilled ceilings etc. can extract heat from a room by providing cool surfaces for heat transfer, and therefore enable cooling to be achieved without full air conditioning, so that in mild conditions cooling can still occur through open windows, etc. This provides a more versatile 'mixed mode' system which is more energy efficient.
- Ice can be used as an effective means of thermal storage, storing 'coolth' in colder parts of the day to provide cooling during warmer parts of the day. Combined with a standard air conditioning system, the ice is produced at night when electricity is cheap, then its thermal capacity is released during the day when using electricity directly for compression cooling would have been expensive.

Ventilation and draughts

Excessive heat gain or loss through air infiltration from the environment into the building can constitute a sizeable percentage of the total energy bill. Reducing draughts is a cost-effective technique and can be accomplished in a number of ways. In dwellings, the application of draught-proofing material round the edges of doors and windows is usually sufficient, while in commercial premises, the use of revolving doors and automatic door closers should also be considered. Calculations showing the effects of insulating buildings, and other interventions are given in Case studies 1 and 2 and Appendix 1.

Building services equipment

The building services equipment uses most of the energy consumed in a building, therefore for maximum effectiveness it should be appropriate to its function, operate efficiently, and be correctly sized. Many systems are significantly oversized and therefore run at much less than optimum performance. It is worthwhile investing in smaller or more efficient items of plant where feasible, as small changes in the rating of a piece of equipment may lead to large reductions in lifetime energy use. For instance, an 11 kW motor costing £700 can consume £67,000 worth of electricity over its lifetime of ten years, thus a small increase in capital cost to acquire a machine of greater efficiency will pay for itself very quickly.

Heating system

Choosing the right heating system has important energy implications. Some systems such as warm-air heating have an output that is almost totally convective, while others are predominantly radiative. Radiators used in central heating systems emit between 30 and 70 per cent of their energy by radiation, depending on the design and the surface temperature. Older systems often have large cast iron radiators, which contain a large volume of water, have a high thermal mass, and therefore exhibit a slow response to changes in demand. More modern radiators, being constructed of thinner metal sheet and also containing less water, are capable of providing a more rapid response to conditions and delivering greater efficiency.

In buildings with high ceilings, a warm-air system may turn out to be very inefficient, since the warm air will rise towards the roof, where it does not benefit the occupants. Destratification fans and ductwork can be fitted to direct the warm air back to ground level and reduce temperature gradients, improving efficiency. Often such spaces are better heated by high-level radiant heaters, which beam radiant heat directly to the occupants and surfaces. Intermittently heated spaces with large amounts of thermal mass may also benefit from high-level radiant heating which heats the occupants directly. An efficient alternative, particularly for housing but also widely used in commercial applications, is underfloor heating, which requires lower water temperatures and can be used effectively with efficient sources of heat such as heat pumps and condensing boilers.

Boilers

Modern conventional gas and oil-burning boilers for heating systems and domestic hot water are very efficient and can reach a maximum efficiency of 75–85 per cent at or near maximum capacity. At lower load fractions, efficiency may drop significantly because of the standing losses. When a large heat load needs to be met, it is usually more efficient to use a number of small boilers, since efficiency falls off at low load fraction, therefore one small boiler operating at a high load fraction is more efficient than a large boiler on low load.

Condensing boilers

Although conventional boilers are now very efficient, even greater efficiency can be achieved by using condensing boilers. When a fuel such as natural gas is burnt, it produces carbon dioxide (CO_2) and water vapour:

$$CH_4 + 2O_2 \rightarrow CO_2 + 2H_2O \text{ in the form of water vapour.}$$

Since the water is in the form of vapour at an elevated temperature, the heat content is significant, and if it is allowed simply to disappear up the flue then it is lost. The latent heat of evaporation and some additional sensible heat can be recovered by using condensing boilers, which are available from domestic scale upwards. After combustion and passing over the main water tubes, the flue gases pass over further heat-exchange surfaces where the water vapour condenses out (see Figure 4.7). Near the top end of the load cycle, the efficiency of such boilers is about 10–15 per cent greater than for a conventional boiler, and even when not operating in the condensing mode, some advantage is obtained due to the greater heat-transfer surface. The lower the return water temperature, the greater the degree of condensation and the higher the efficiency of the boiler, as condensation can only take place when the return water is at a temperature less than about 59°C.

Table 4.1 Typical efficiency of a condensing boiler at a range of return temperatures

Mode	Efficiency %
Non-condensing	85
Return temp 40°C	90
Return tem 30°C	95

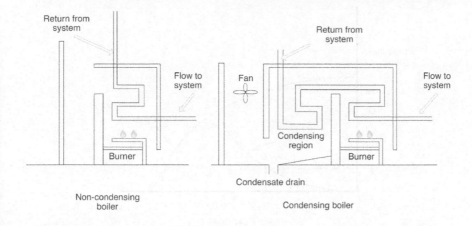

Figure 4.7 Non-condensing and condensing boilers

Condensing is particularly effective for a system with a low water return temperature such as underfloor heating. There is some increase in installation costs, as a 250 kW boiler with a 40°C return water temperature produces 14 litres of condensate per hour and a drain is therefore required to remove it. As the condensate is slightly acidic, the drain should be of plastic, and stainless steel flues are preferred since copper and cast iron are susceptible to corrosion. A fan is normally required to remove the flue gases, which are cooler (at 55–100°C) and therefore of lower buoyancy than in a conventional boiler. The cost is 30–50 per cent higher than a conventional boiler and the payback period is two to five years.

Variable speed drives (VSD) for pumps and fans

Pumps consume about 12 per cent of the world's electrical energy, and most of them consume over 50 per cent more than they would if they were designed, specified and used correctly. The energy use over its life cycle is many times the cost of a pump, so it is worthwhile investing in more efficient pumps and control devices. Older heating, ventilation and air-conditioning (HVAC) systems often have only single-speed fans and pumps, while newer ones may include two-or three-speed plant. Since pumps and fans need to be sized to meet the maximum demand, for a substantial proportion of their life the load is only a fraction of this and energy is wasted if a conventional regulator or damper system is used. Figure 4.8 shows the operation of a fan in a ventilation system. Where the fan curve and the system curve cross (A) is the operating point, and the power consumed by the fan is given by the area ADOG. When demand falls, the airflow rate is reduced by closing a damper, resulting in

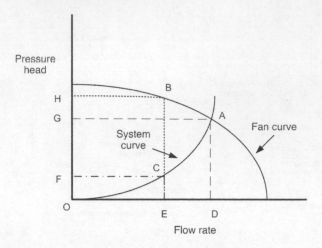

Figure 4.8 Operation of a fan in a ventilation system

increased pressure loss and reduced flow, and shifting the operating point to B. The power consumption is now given by the area of rectangle BEOH, which is little lower than the original value, although the flow rate has fallen by almost 50 per cent. An efficient alternative way of restricting the flow is to use an inverter drive, which allows the flow rate to be reduced by slowing down the fan, resulting in a corresponding lowering of the energy input; thus the power consumption at flow rate E is given by the area CEOF, around one quarter of the value obtained by using dampers. Annual fan and pump energy savings of over 60 per cent are claimed by users of VSDs, which in many cases can be cost-effectively retrofitted to existing systems.

Sample calculation

For pump efficiency of 80 per cent and VSD efficiency of 87 per cent, the flow rates, pressure drops and power consumption at points A, B and C in Figure 4.8 are as follows:

	Flow m^3/s	Pressure Pa (Pascals)	Power
A	8	220	2200
B	6	230	1725
C	6	85	732

While the flow rate has fallen by 25 per cent, in case B the energy consumption has fallen by only 21 per cent but with the VSD the consumption falls by 67 per cent.

Domestic hot water

In homes, the provision of hot water can consume up to 25 per cent of the overall energy demand and almost half of the heating requirement; in a well-insulated building it may even constitute the dominant heat load. In offices, the proportion is lower, but when refurbishing it may often be more efficient to use point-of-use water heaters, even if electrically heated, as this avoids the need to use the boiler during the summer months.

Heat pumps

A heat pump employing a compression refrigeration cycle is the usual way of providing cooling to an air conditioning system, a heat pump used in reverse mode can upgrade low-level heat from the air, the ground or running water, and also release it at a higher temperature to provide space heating in a building. As it is electrically driven, it is therefore useful where there is no mains gas supply. Heat pumps are supported by the British government under the renewable heat incentive (RHI). Their advantages include reduced servicing, no boiler, high security, long life, no flue or ventilation and no local pollution. In appropriate situations they can also save money and reduce carbon emissions.

The efficiency of the system is defined by the coefficient of performance (COP), which may have different values for heating (COP_h) and cooling (COP_c). Typical values range from 3–4 for COP_h, and 2.8–3.6 for COP_c. They are defined as:

$$COP_h = \frac{Evaporator\ duty}{Compressor\ power}$$

$$COP_c = \frac{Condenser\ duty}{Compressor\ power}$$

In other words, the amount of heat transferred is roughly three times the electrical energy input. The COP falls as the temperature difference between the evaporator and condenser increases. At large temperature differences, supplementary heating may be needed, adding to the capital costs.

Heating mode

Air-source heat pump

Air-source heat pumps (ASHP) have the advantage of relative simplicity, and are therefore the easiest option to specify for retrofit projects. A fan blows air over the evaporator coil located on an outside wall of the building (see Figure 4.9). Installation of ASHP is relatively cheap but they suffer

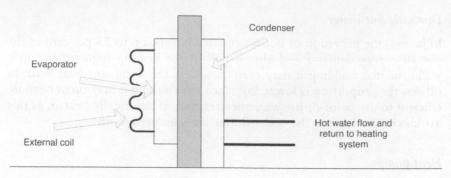

Figure 4.9 Air-source heat pump

from the drawback that the temperature of the air is at its lowest when the greatest amount of heat is required, leading to low values of COP at high load. Also, as heat is extracted from the external coil, regular defrosting is necessary. ASHPs are therefore of most benefit in moderate climates. Typically, an air-source heat pump can provide 85 per cent of the heat for a building, and a conventional boiler is used to provide the rest. ASHPs have a typical COP of 4.4 at an air temperature of 7°C and a flow temperature of 35°C.

Use of water as heat source

Closed-loop system

Because of its high thermal capacity, the temperature of a body of water varies much less than that of the air above it. As a result, the large falling-off in COP experienced in air-source systems in severe conditions is much reduced. Sea water in estuaries and water in rivers has been used successfully in this way. In these systems water is used as the circulating heat-transfer medium, passing through a heat exchanger on the evaporator side and a coil placed in the water.

Open-loop system

An alternative to a circulating system is an open-loop system, which uses water that is pumped from a deep borehole and passes over the evaporator coil, after which it is led to a soakaway. The temperature of the water is close to that of the ground, which in the UK is between about 9 and 12°C; it is therefore often used for low-cost cooling (Figure 4.10). The running cost of this is relatively low because the only energy input is that required to pump the water from the ground. The effective coefficient of system performance (COSP) lies between 10 and 100.

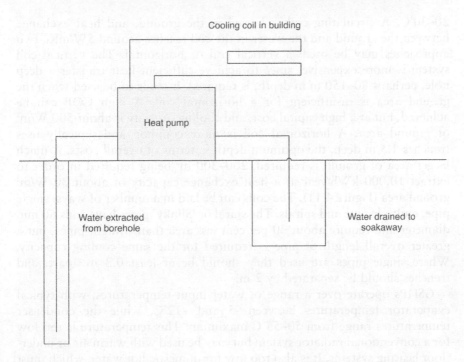

Figure 4.10 Open-loop system

A licence to extract water is normally required from the Environmental Protection Agency (EPA) in England and Wales or the Scottish Environmental Protection Agency (SEPA) in Scotland. The agency may also insist on additional requirements such as returning the water to the aquifer, and there are also charges (currently £10–25/litre in the UK), which may affect the feasibility of a scheme. Water quality should be high, with low dissolved solids, and the available flow rate should be 25–50 l/s. The additional capital costs may be considerable, and include the drilling of test boreholes.

Some notable large developments have used groundwater in association with heat pumps or other systems, including Portcullis House in London, an office block for MPs. The Scottish Parliament building in Edinburgh uses water from a spring for 'free' cooling and expels the water to ponds in the grounds.

Ground source heat pumps (GSHP)

The ground a metre or more below the surface of the UK has a relatively stable average temperature, ranging from 9°C in the far north to 12°C in the south, while in hot climates such as India and the Far East it may reach

20–30°C. A circulating coil is buried in the ground, and heat exchange between the ground and the water in the coil reaches around 5W/mK. Two approaches may be used: a vertical coil or horizontal. The vertical coil system is more expensive, since to achieve sufficient heat transfer a deep hole, perhaps 80–150 m in depth, is required. It tends to be used when the ground area is insufficient for a horizontal coil. A high COP can be achieved, but at a high capital cost, and cooling capacity is about 500 W/m^2 of ground area. A horizontal coil is more common and typically uses trenches 1.5 m deep, the optimum depth in terms of overall costs. A much larger area of ground is required, 200–300 m^2 being required in order to extract 10,000 kWh/year at a heat exchange capacity of about 30 W/m^2 ground area (Figure 4.11). The coils can be laid in a number of ways: single pipe, multiple pipe and spirals. The spiral or 'Slinky' pipes have coils 60 mm diameter and require about 30 per cent less area than straight pipes, but a greater overall length of pipe is required for the same cooling capacity. Where single pipes are used they should be at least 0.3 m apart, and trenches should be separated by 2 m.

GSHPs operate over a range of water input temperatures, with typical evaporator temperatures between -5 and +12°C, while the condenser temperatures range from 50–55°C maximum. This temperature is too low for a conventional radiator system but may be used with warm air or under-floor heating systems. It is also too low for domestic hot water, which must

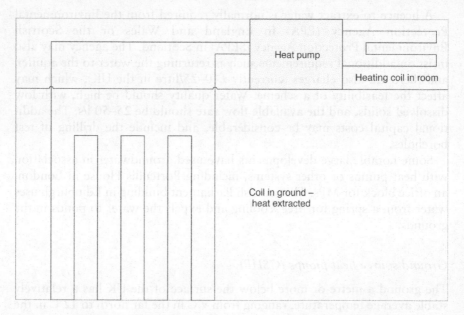

Figure 4.11 Ground source heat pump

be at a minimum of 60°C to reduce the risk of legionella. A supplementary heater, usually electric, can be used after pre-heating the water to 50°C with the heat pump. Problems may arise if the fluid in the ground loop is below 0°C, as freezing may cause ground heave. Typical COP values for a GSHP in UK conditions are shown in Table 4.2. COPs of 3.5 and over give a significant advantage over gas-fired heating. On average a GSHP produces 65 per cent less CO_2 than oil-fired heating, and 45 per cent less than gas.

Design considerations

A heat pump may be designed to cover full or part load. Part-load systems significantly reduce the capital cost and would normally be sized to cover about 50 per cent of the load under winter design conditions. On this basis, for most of the year the pump would provide 90–95 per cent of the heating requirements. Top-up heating to cover the lowest temperatures may be provided by low-capital-cost electric heaters or, particularly for dwellings, wood-burning stoves. The source of heat and the type of heat distribution should be matched according to temperature requirements (Table 4.3).

Domestic heat pump use

Although heat pumps are becoming more popular in the UK, their use is not as widespread as in Scandinavia and some other regions. Packaged units which combine an air-source heat pump and top-up heating, together with controls, are available, and may also include domestic hot water and cooling options. Such systems may also be used in combination with fossil-fuel boilers (connected in series to the heat pump) where the heat pump provides pre-heated water to the boiler for the coldest periods. For refurbishment projects the boiler may also be connected in parallel with a heat pump of lower capacity so that the boiler covers the full load at extreme temperatures. Whether the fossil-fuel boiler or electric heat pump is the most cost-efficient supplier of heat is determined by the relative fuel costs. Such a system provides the flexibility to use the cheapest fuel option at the time.

Table 4.2 Typical heating Coefficient of Performance (COP_h) for a range of temperature outputs

Water output temperature °C	COP_h
55	2.4
45	3.2
35	4.0

Table 4.3 Temperatures required for heat distribution systems

Heating system type	Temperature °C
Underfloor	30–45
Low–temperature radiators	45–55
Conventional radiators	60–90
Warm air	30–50

Controls for heat pumps

The most energy-efficient form of control uses a variable speed drive to the compressor via an inverter. In this way the output of the pump can be matched to the demands of the building, and as the required heat output falls, so does the energy input to the compressor.

Performance comparison

A comparison between the performance of different technologies is shown in Tables 4.4 and 4.5 for heating and cooling respectively.

Table 4.4 Comparison of heating performance of condensing boiler, air source heat pump and ground source heat pump

	Condensing boiler	Air source heat pump	Ground source heat pump
Heating load (kW)	40	40	40
Seasonal efficiency (%)	88	300	400
Energy input (kW)	45.4	13	10
Cost of energy (p/kWh)	4.5	11.0	11.0
Hours run	2000	2000	2000
Cost of input energy (£)	4086	2860	2200
CO_2 emitted (kg)	21.6	14.2	10.75

Table 4.5 Comparison of cooling performance of air source heat pump and ground source heat pump

	Air source heat pump	Ground source heat pump
Cooling load (kW)	40	40
Seasonal efficiency (%)	280	320
Energy input (kW)	35.7	31.25
Cost of energy (p/kWh)	11.0	11.0
Hours run	500	500
Cost of input energy (£)	1963	1718
CO_2 emitted (kg)	15.3	13.2

Mechanical ventilation and waste heat recovery plant

Natural ventilation is obviously a lower energy-cost option than mechanical ventilation but in cases where mechanical ventilation is unavoidable, fan motors should be fitted with VSD as described earlier in the chapter. In housing in the UK, ventilation is usually by natural means and we try to minimize the ventilation and infiltration rates in winter in order to keep heating loads down. However, when mechanical ventilation is used, then energy costs can be kept low by recovering heat from the waste ventilation air. Mechanical ventilation with heat recovery (MVHR) has been used successfully as a retrofit in 'problem' housing to reduce excessive moisture, combating condensation and mould growth, thus saving money on maintenance and decoration in the long term. MVHR is also used in the 'Passivhaus' approach to housing design, where the ethos is to produce a highly insulated, well-sealed house, and provide constant mechanical ventilation with heat recovery when required in winter.

Devices used for heat recovery include plate heat exchangers, run-around coils, and thermal wheels. In ventilation applications, air–air plate heat exchangers are frequently used where the inlet and outlet are in close proximity to each other, and thermal wheels have been used successfully in many applications.

In ideal situations, the effectiveness of all types of heat exchanger may exceed 80 per cent, but the values realised in practice tend to be much lower and depend largely on the operating conditions. Pressure drops within the heat exchangers should not be ignored as they add to the running costs and may reach 100–200 Pa in a thermal wheel and up to 250 Pa in run-around coils. Where there is a substantial distance between inlet and outlet, run-around coils are more appropriate. The energy costs of ventilation can be reduced by using demand-controlled ventilation, based on room CO_2 levels, combined with the use of inverters to control variable-speed fans.

Combined heat and power (CHP)

A CHP system is a heat engine that produces electricity through a generator and is able to make use of the rejected heat from the engine in a water- or space-heating application. Many small CHP units use diesel or gas power and have similar reliability factors to conventional boilers; availability factors of 95 per cent and above are common. The great advantage of CHP is that overall energy efficiency is in the region of 80–90 per cent, as shown in Figure 4.12. A typical energy balance is shown.

Low-grade heat refers to heat emitted at 35–45°C, while high-grade heat has temperatures in the region of 70–85°C. The usefulness of a CHP system depends on a number of variables: a simultaneous demand for heat and electricity, the ratio of heat to electrical load, the capital cost, the fuel costs for the system, and the cost of the alternative (usually mains) electricity and

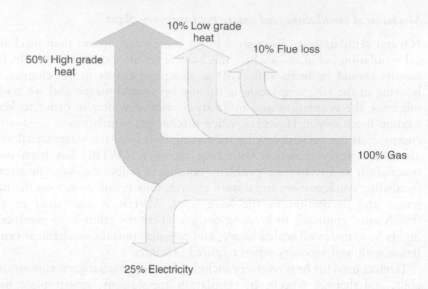

Figure 4.12 CHP system energy balance

heat. Viability depends to a large extent on the difference in utility rates. To make CHP worthwhile, the cost of electricity should be at least three times the cost of heat.

General CHP considerations

In general, a CHP system needs to operate for at least 4,000 hours per year to be cost-effective. If well-designed and matched to the loads, it can give a payback period (PBP) of three to five years. Figure 4.13 shows a typical CHP system which consists of an engine or turbine, a generator, a heat recovery system, an exhaust system and a control system.

CHP systems are defined as micro (<5 kW), mini (5–30 kW), small (up to 5 MW), medium (5–50 MW) and large (>50 MW). Larger applications may use gas turbines or steam systems, while Stirling engines are used only for domestic-scale systems where electricity production is low. Many small-scale units having an electrical output up to 1 MW use automotive-derived engines altered to run on gas. The heat to power ratio is typically as follows:

    up to 3:1 for microturbines
    up to 2:1 for engines
    1:1 for fuel cells.

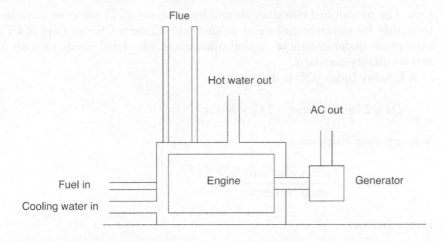

Figure 4.13 Typical CHP system

Dual-fuel engines tend not to be used as they are not optimized for efficiency for either fuel, therefore their potential CHP efficiency is low. Most of the heat (50 per cent) is recovered from the engine cooling jacket and the exhaust system in the form of low-temperature hot water (LTHW), with low-grade heat being the remainder (10 per cent). The heat from the exhaust gases is recovered using a gas-to-water heat exchanger with the heating circuit water flowing directly through it. A further condensing heat exchanger may be added to remove latent heat from the flue gases. A buffer tank may be installed between the CHP and the hot water return to reduce frequent cycling. As connections to the mains will normally be maintained, agreements have to be made with the local distribution network organization and appropriate control gear installed. If the electricity generation is less than 16A per phase, a less formal arrangement is required but the distributor must be informed.

Criteria for installing CHP

A reasonably large heat load should exist over the greater proportion of the year, to enable the plant to run for 11 hours per day for the whole year or 17 hours per day for eight months of the year. Heat and electrical loads should be simultaneous, space should be available for the unit, and the electrical load should remain above the output of the unit for most of the operating hours. Mini CHP should be sized to match the base heating load to maximize running hours, and in a mixed system the CHP should be the lead heating appliance. If the CHP is replacing gas for heating, then any gas used for cooling should be discounted, as it cannot be replaced by CHP heat. The overall fuel efficiency should be in excess of 75 per cent. In order to qualify for government incentives such as a Climate Change Levy (CCL) exemption under enhanced capital allowances, the CHP needs to meet a certain quality standard.

A Quality Index (QI) is defined as:

$$QI = 249 \times hpower + 115 \times hheat$$

Where power efficiency

$$hpower = \frac{total\ power\ output}{total\ fuel\ input}$$

Heat efficiency

$$hheat = \frac{qualifying\ heat\ output}{total\ fuel\ input}$$

The constants 249 and 115 are related to the alternative electricity supply and alternative heat supply options displaced by the CHP. Qualifying heat output is the amount of useful heat supplied annually from a scheme that can be directly shown to displace heat that would otherwise be supplied from other sources. QI should be at least 105 with a minimum power efficiency of 28 per cent to qualify, and metering is required for input and both outputs.

Example of CHP calculation for residences at a university campus in the UK

In this example, heating is provided by gas, and cheap off-peak electricity is available at night. It is generally not cost-effective to run the unit during the night because of this lower cost. For the campus, mains natural gas is the obvious fuel for the CHP as it is already available at a favourable tariff.

Without CHP, current energy use for residences is:

	Consumption kWh	Cost £
Electricity @ 11p/kWh	874,564	96,202
Gas @ 4.5p/kWh	2,679,311	120,568

taking existing boiler efficiency as 80 per cent and with an estimated load factor of 75 per cent. For year-round base load operation, we need the hourly useful heat load (HL) averaged over the summer months.

$$HL(kWh) = \frac{(H_{June} + H_{July} + H_{August}) \times Boiler\ efficiency}{No.\ of\ days \times Hours\ per\ day}$$

The consumption for these months is taken from the billed consumption data:

$$HL = \frac{(116,728 + 140,118 + 120,036) \times 0.8}{89 \times 20}$$

$$EL = \frac{(E_{June} + E_{July} + E_{August}) \times F}{89 \times 20}$$

where F is utilization and is 1 for separate day and night metering, 0.85 for flat-rate metering.

$$EL = \frac{(53,832 + 58,335 + 50,026) \times F}{1,780}$$

$$EL = 77.45kW$$

Therefore, average daytime low electrical demand is 77.45 kW. If no electricity was exported, then the CHP unit would need to have an electrical output of no more than 77.45 kW. The electricity cost for this period = £17,841 and gas cost = £16,962.

Selecting from a manufacturer's table of outputs, unit A below would be suitable, having an electrical output of 77 kWe, a corresponding heat output of 123 kW and a gas input of 233 kW. Savings are worked out as follows:

Unit A: savings p/hour, total p/hour

Savings	p/hour	Total p/hour
Electricity 75 × 11	825	1,528
Gas 125/0.8 × 4.5	703	
Costs		
CHP gas 233 × 4.5	1,048.5	1,101
Maintenance @ 0.7p/kWhe	52.5	
Electricity generated = 0.75 ×75		
Net benefit		427

The saving is £4.27 for every hour of CHP operation. Therefore the annual number of operating hours can be given by the load factor × total hours per year = 0.8 × 8,760 = 7,008 hours, minus 5 days per year for servicing = 6,888 hours. Therefore the potential hours of operation are 6,888, multiplied by 4.27 giving annual savings of £29,411. Since the electricity could be exported, a base heat load sizing method could be used. Unit B, with an electrical output of 85 kW and a corresponding heat output of 143 kW, and gas input of 254 kW could be used.

Unit B: savings p/hour, total p/hour

Savings	p/hour	Total p/hour
Electricity 85 × 11	935	1739
Gas 143/0.8 × 4.5	804	
Costs		
CHP gas 254 × 4.5	1,143	1,206.75
Maintenance @ 0.7p/kWhe	63.75	
Electricity generated = 0.75 ×85		
Net benefit		533.75

Net saving is £5.33 for every hour of CHP operation, and the annual potential savings are £36,713, not taking into account feed-in tariffs or other financial incentives.

The CO_2 savings from such a scheme can also be estimated. The CO_2 saved per hour from electricity generation by the CHP unit, as opposed to buying it from the grid, equals 70.55 kg. There is, however, an increase in the use of gas. The existing CO_2 emission for gas equals 35.75 kg/hour, and the new CO_2 emissions are 50.8 kg/hour. Therefore there is an increase of 15.05 kg/hour. The net benefit is therefore 70.55–15.05 = 55.5 kg of CO_2 saved per hour. This results in an annual saving of 382 tonnes of CO_2 per year.

Combined heat and power summary

The general design procedure is to identify the heat demand profile and the electricity demand profile, and size according to the heat load. Heat dumping should be avoided and can be obviated by the use of thermal storage (typically a large insulated tank), which enables electricity sales to be maximized. The inclusion of a large number of diverse users affords opportunities to even out loads over time, thus making for greater efficiency.

Air-conditioning

Where there is an existing air conditioning (a/c) system, significant improvements may be difficult to effect, but regular servicing and maintenance should be implemented to achieve optimum performance. Many a/c plants are oversized to cope with the hottest conceivable day, with some capacity to spare. As a result, compressors run most of their lives well below full load and with low efficiency. A refurbishment may offer opportunities to replace some items of plant with smaller, more efficient units. Changes to the building fabric and operation may reduce some of the plant loads significantly to allow this. Night ventilation to reduce the daytime cooling load by reducing building fabric temperatures is often incorporated into the design strategy of new buildings and may be implemented in an existing building. The feasibility of doing so will depend on a number of factors, such as the structure and layout of the building, the expected outdoor temperatures, and the daytime heat gains in the building. In a heavyweight building where night temperatures fall well below 20°C and there is a cooling load during the day, then night cooling should be seriously considered. It is essential that there are sufficient internal openings to allow the ventilation air to flood the building, but security should not be compromised. If daytime temperatures exceed 36°C then night ventilation is unlikely to have a significant effect. Among the possibilities for improving energy efficiency is the replacement of mechanical ventilation and a/c with natural ventilation. To assess the viability of natural ventilation the following considerations should be borne in mind.

Environmental

If peak summer heat gains are in excess of 40–50 W/m² then mechanical ventilation or air-conditioning are likely to be necessary. In urban locations, noise from roads or railways nearby may act as a serious deterrent to the use of natural ventilation. The noise level should be at most 70 dB at the façade to achieve an acceptable level of 50–55 dB with windows open. Airborne pollution from traffic or industry may similarly render natural ventilation impracticable. The presence of a prevailing wind should be noted, as it may affect the strategy adopted. The effect of surrounding buildings, with respect to light, wind movement and sound, should also be considered.

Building

Shallow-plan buildings are easier to naturally ventilate than deep-plan. For single-sided ventilation, the maximum depth is 7–10 m, for double-sided cross-ventilation (open plan) up to 15 m is possible. Where mechanical ventilation is required, UK Building Regulations specify minimum efficiency levels for fans, i.e. maximum specific fan power (SFP), which is the number of watts needed to move one litre of air per second. For new domestic buildings this ranges from 0.5 $Wl^{-1}s^{-1}$ to 1.5$Wl^{-1}s^{-1}$ and for non-domestic buildings 0.3 $Wl^{-1}s^{-1}$ to 1.9$Wl^{-1}s^{-1}$, depending on the particular application within the building. Electrically commutated (EC) motors are more efficient, and can give a SFP of 0.3 compared with 0.8 in a conventional motor, and speed can be controlled better and more efficiently.

Lighting

Lighting accounts for a significant proportion of energy costs in commercial buildings and therefore presents huge opportunities for savings. Although heating is the major user of fuel, because of the cost differential between heating fuels and electricity, lighting may represent 30–50 per cent of the total energy bill. When a room is too warm we notice very quickly, and make adjustments to the temperature, but we do not necessarily notice when there is too much light, as the eye can adapt over a wide range of lighting levels. Because of this, lighting is often poorly controlled. Another source of waste is the problem of glare, where direct sunlight through a window makes it difficult to read or use a computer screen; a common solution is 'blinds down, lights on', and often the blinds remain down when the sun has gone, and lights remain on when not needed. Lighting and shading provision need to be coordinated in order to provide visual comfort and energy efficiency.

Energy can be saved in lighting in a number of ways – by increasing its efficiency, by reducing the level of lighting, and by cutting the hours of use. The use of compact fluorescent lights and LEDs is now widespread in homes; lighting energy represents only a small proportion of the annual fuel bill and manual control is still the norm. In commercial premises, although the use of efficient fluorescent tubes is long standing and LEDs are making serious inroads, the need for good control of lighting remains.

Lighting control options

Manual control:

- position switches in a convenient location so as to encourage users to switch off
- group luminaires and switches so that rows of lights close to windows can be switched off to take advantage of good levels of daylight
- colour-code groups of lights in buildings such as supermarkets for different sets of users who require different lighting levels, such as cleaners and shelf-stackers. High levels of lighting are usually only required when customers are in the store.

Timed control:

- switch lights off at natural break times and at the end of the working day. Allow manual override of main lights or use locally-controlled task lighting.

Sensor control:

- use occupancy-detection sensors such as passive infra-red (PIR) or ultrasonic
- use daylight sensors to automatically switch groups of luminaires off when daylighting is adequate
- use daylight sensors in conjunction with dimmers to allow electric lighting to top up daylighting
- use individually addressable luminaires for greater flexibility.

Escalators and lifts

In certain building types, such as department stores and shopping malls, escalators may consume up to 15 per cent of the total energy used in the building. In such buildings, escalators are often running continually during opening hours, and in some cases considerably beyond. The first priority should be to ensure that they do not run beyond the hours required. This is easily achievable with a time switch. Further savings can be achieved by more sophisticated control using occupancy sensors. For instance, when there is no demand the speed can be reduced, or the escalator stopped altogether. Speed reduction can be achieved using star-delta control or variable voltage, variable frequency drive (VVVFD). The greatest energy savings are achieved by stop/start control followed by VVVFD, but it results in more wear to the moving parts and increased maintenance costs. Lifts can also benefit from VVVFD which enables energy consumption to be reduced through speed reduction.

Server rooms

These are now widespread and are responsible for much of the energy used in specific building types, particularly those operated by banks and other financial institutions. Data centres now comprise a significant portion of the cooling load of many commercial buildings, and they are now thought to constitute about 3 per cent of the UK energy load. It may be possible to arrange for certain operations in server rooms, such as routine back-ups, to be carried out at the times when the electricity tariff is lowest.

Energy supply and the use of renewable energy in buildings

Changing suppliers

Many countries have energy supplies provided by the private sector, and since the privatisation of electricity and gas utilities in the UK it has been possible for consumers to change suppliers in order to save energy. Often there are 'one-off' incentives to change, which are particularly attractive to domestic consumers, as is the possibility of having only one supplier (and one bill) for both gas and electricity supplies. Commercial organisations have been more reluctant to take advantage of these changes, due partly to 'inertia', and to some extent to a lack of understanding of the costs of utilities and the benefits of changing supplier, and concerns that the process may bring problems with it, and even disruption to supplies.

A good general principle to follow to reduce energy use and emissions is that the load should first be reduced as far as possible, then a carbon-efficient means of meeting the load found, whether that is from a fossil or renewable source. There are opportunities to use renewable energy in buildings to reduce reliance on fossil fuels. The feasibility of the different technologies is a function of climate, capital cost, and the cost of fuel replaced, bearing in mind that for many systems such as photovoltaics (PV) and wind turbines it is necessary to maintain a mains connection and incorporate an inverter and control unit. Depending on location and local economics and fuel costs, some of the following may be appropriate.

Suitable systems include solar water/air heating, solar electricity (PV) and building-integrated PV, solar refrigeration, and roof-mounted wind turbines. It is considered by many that micro wind turbines are unlikely to pay back either their carbon emission or costs, and vibration effects related to building-mounted turbines suggest that larger turbines sited more remotely are the future for wind power. Transmission losses from such locations are not great – only 10 per cent losses are incurred if electricity is transmitted the whole length of the UK. PV systems are not so sensitive to scale factors, but it remains the case that the control gear for a very small PV system costs not much less than that for a large one. Building-integrated PV offers many opportunities for retrofit applications.

Most renewable electricity supplies will be linked via an inverter and control box which continues the mains supply to the building, as it is unlikely that renewables will provide all the power needed throughout the day and year, and it is also important to the economic viability of a system to be able to sell excess electricity back to the grid. Control of the electricity supply conditions is important, as varying voltage, phase and power factor can have an impact on the energy efficiency. Where the power factor diverges substantially from a value of 1 the apparent power is reduced, and this occurs where there are large inductive loads. Power-factor correction via a bank of capacitors is normally used to bring the power factor closer to 1. Often the voltage delivered by the electricity supplier is higher than that needed to run the plant or appliances, and energy can be saved by optimising the voltage. It may also help to prolong the life of the plant by reducing excess energy which is ultimately rejected as unwanted heat or vibration. Voltage optimisers may also improve the quality of the supply by filtering out the harmonics and transients which sometimes result when power is fed in from renewable sources.

A side effect of the increase in generation from renewables such as PV is the rise in DC power generation, particularly for domestic premises. In most systems this is converted to AC through an inverter and fed into the grid supply system to either power devices or fed back into the grid. The conversion process naturally results in energy losses, although new amorphous low-loss transformers are available that reduce them. It can be argued that using the DC directly would be beneficial because many devices such as phones, laptops etc., need DC chargers; also some other devices can operate more efficiently using DC. Fan-coil units operating on low-voltage DC are typically 50 per cent more efficient than AC versions. Other low-voltage DC appliances are being developed, but there are considerable difficulties involved, particularly as both AC and DC supplies would need to be maintained and the economic viability would need to be carefully investigated.

Summary

A number of measures can be implemented to improve any building. The cost-effectiveness of each measure will vary with location and with the individual building. A rough order of effectiveness is given in Figure 4.14 (in approximate order of decreasing cost effectiveness based on cost per kg CO_2 saved). The measures applicable to specific building types will vary greatly depending on the use and the hours of operation; some examples of potential savings in shopping centres are shown in Table 4.6.

Table 4.6 Potential energy savings in retail shopping centres

Measure	Potential energy saving %
Switch off unessential lights outside opening hours using timers	10
Switch off unessential lights in back-of-house areas	5
Replace tungsten bulbs with compact fluorescent bulbs	75
Install photocell controls to make use of daylight; use presence detectors in areas such as toilets	20
Switch off escalators outside running hours	15
Link external car parking lights to daylight	60
Optimise switch-off of HVAC system	20
Use night cooling to reduce air-conditioning	20
Set thermostats correctly	1°C = 7%
Check controls so not heating and cooling at same time	10

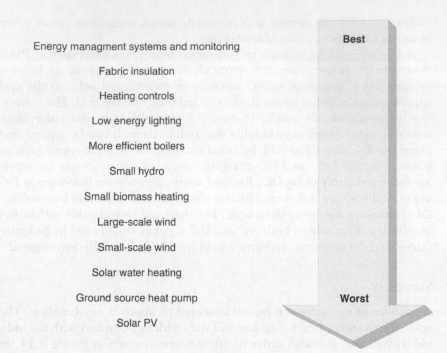

Figure 4.14 Cost-effectiveness of energy saving measures

5 Instrumentation and measurements

It is an old adage of management that 'if you can't measure it, you can't manage it', and while information from energy bills can be used in the energy audit, it may not provide a sufficiently fine level of detail and it is often necessary to have additional information taken from measurements of the conditions in the building, the state of the plant, and the energy consumption. Additional measurements can be carried out to evaluate the performance of individual items of plant. Monitoring a building over a short or even an extended period may be considered necessary in order to elicit this information. The equipment required for this purpose includes handheld instruments, data loggers and a range of sensors. A large range of instruments is now on the market and the measurements required for energy management can be made relatively easily; modern electronic instruments are compact, durable, and inexpensive. The main measurements of interest are:

- fuel consumption
- temperature
- electrical power
- ventilation and air movement
- water flow
- relative humidity
- heating efficiency
- U-value.

Fuel consumption

Fuel consumption may be measured in cubic metres of gas, litres of fuel oil, etc.; conversion to kWh and carbon dioxide emissions may be carried out using Table A4 in Appendix 2. For meaningful comparison with other buildings of the same type, the consumption should be calculated on the basis of square metres of treated floor area. Used in conjunction with internal and ambient temperature readings, degree-day information can be produced to enable the kind of analysis shown in Chapter 2 and also to enable CUSUM analyses. Where a building management system (BMS) is used, it can be configured to collect and store metered data, and with the appropriate software options, degree days and other data can be generated.

Temperature

The most important measurements, and the most numerous in building applications, are measurements of temperature. Typical measurements required include internal and external air temperature, radiant and globe temperature, surface temperature, and fluid temperatures in pipes and ducts. The range of useful sensors for data loggers and BMS is limited to those with an electrical output such as thermocouples, thermistors and resistance thermometers. For on-the-spot measurements, a number of inexpensive handheld instruments are available, many of which incorporate humidity sensors and may also have facilities for linking to data loggers.

The temperature measured by a globe thermometer is generally considered to give a response closely related to that of humans, and combines the effects of air and radiant temperatures (the operative or dry resultant temperature). The instrument comprises a temperature sensor located at the centre of a 40 mm diameter sphere whose surface is painted matt black.

All bodies at temperatures above absolute zero emit infrared (I-R) radiation in the wavelength region 700nm to 1mm, allowing the temperature of a body to be measured as a function of the thermal radiation it emits. I-R thermometers allow the surface temperature of a body to be measured using a non-contact method, having the advantage that measurements may be made from some distance away from the surface. I-R cameras are regularly used in thermographic surveys, in which they produce an image of the object being investigated, which is colour-coded according to temperature. By surveying the outside of a building, preferably at night in winter, it is possible to locate places where excessive heat is being lost, for instance where insulation is missing or where there is a thermal bridge. Small handheld I-R cameras are now available inexpensively for purchase or for hire. Considerable expertise is needed to avoid misinterpretation of the results, however. For example, an outer wall displaying a low temperature may indicate low heat loss due to good insulation, but it may also occur simply because the room behind it is unheated.

Care is needed to ensure that any sensor used is measuring the desired quality. A sensor intended to measure the ambient air temperature should not be exposed to direct solar radiation, as it will absorb energy and measure not just the air temperature but a combination of air and radiant temperatures, which could be more than 10°C higher. To measure air temperature, direct radiation should be excluded by placing the sensor inside a radiation shield, which can be as simple as a polished metal cylinder surrounding the sensor, and large enough to allow free passage of air around the sensor. Alternatively the sensor should be mounted in a location where it is always in the shade. Further errors may be introduced into readings as a result of the response time of the sensor or instrument. However, in general, energy-related variables in buildings change slowly compared with the speed of response of modern instruments. Temperature changes within a room occur at a rate of fractions of a degree per minute, while sensors take only a few milliseconds to respond.

Electrical power

Electrical energy is used for lighting, small power-load equipment (such as laptops and phone chargers, copiers, printers and home entertainment units), larger-load items such as fans, pumps, immersion heaters, lifts and escalators, and industrial equipment including machine tools and compressed-air machines. While the overall electricity consumption of a building or site will be metered, it is unusual to find separate metering for individual circuits. The question needs to be asked: does the data enable you to identify the energy flows with sufficient accuracy and detail or does it point to the need for more sub-metering?

To isolate a circuit or individual item of plant or appliance, a current clamp can be used; the current induced in the clamp is proportional to that in the primary circuit and can be fed to a data logger. For items drawing constant power, hours-run meters can be used, although they give no information on the timing of the power consumption. Sub-metering may be installed during a retrofit or as part of a BMS installation, and while it may be expensive to retrofit complete meter systems, there are circuit-level systems available that can collect power-usage data from individual circuits. Thus, the lighting circuit can be isolated, as can ring circuits for individual floor levels. The areas of high consumption can thus be identified more readily, at a lower cost than full sub-metering.

Smart meters can provide real-time data on energy consumption, and are particularly useful in homes, where a better understanding of how we use energy can often help householders to reduce their use. There may be opportunities for smart meters to interact with powered devices and also with the supply grid upstream to control energy use.

In the UK, the Energy Act 2008 required domestic dwelling to have smart meters for electrical power and gas installed by 2020. The intention was also to provide them for small and medium sized non-domestic buildings too.

One advantage of a smart meter is the ability to transmit near-real-time data to the energy utility. This transmission can take place in a number of ways: through the internet; by using GPRS technology (general packet radio service); or by PLC (power line carrier). For gas supply, transmitting data can be more problematic as there is not always electrical power available at the meter.

AMR (automatic meter reading) has been around for some time but in itself only allows for transmission of energy data, with no other functions.

It brings with it certain advantages, including removing the need for meter readers. Automatic meter management technology (AMM) can carry out additional functions such as providing real-time information to the user, remote billing functions, enabling the linking of other devices, and the ability to control electrical load for some devices using remote switching. This could include scheduling washing machines, charging electrical vehicles, and other appliances to be used in off-peak periods to minimise the cost of the electricity used by the consumer.

Stand-alone devices are readily available allowing the householder to display real-time consumption data. Some are capable of communicating with a smart meter, others work independently and may be able to display data on a TV or laptop.

In the long term, many more small-scale sources of electricity generation will be coming on-stream, such as PV arrays and CHP systems, requiring more sophisticated and active network management. A number of issues remain to be addressed regarding the implementation of smart metering, such as data security and cost effectiveness.

Ventilation and air movement

Measurements of air movement may be made for a number of reasons. They include:

- local air movement in rooms
- movement of air through ducts
- ventilation or infiltration rates.

Measurement of local air movement

Qualitative assessments of air movement, draughts or leaks can be made using plumbers' smoke pellets, while local air flows within rooms are usually too low to make any accurate measurements of air velocity.

Movement of air through ducts

Ventilation rates in mechanically ventilated buildings can be assessed by measuring the rate of airflow through the ducts. The range of instruments includes:

Rotating vane anemometer

These are simple and relatively robust; a small vane assembly is coupled to a mechanical counter via gears. As this is a mechanical device the lower limit of measurement (0.5 m/s) is determined by friction in the bearings and gears.

Hot wire anemometer

The sensing head comprises a wire which is heated by an electric current, and when placed in the airstream a cooling effect is exerted on the wire by convection. As the air speed increases, it leads to an increase in the convection heat-transfer coefficient and greater cooling. The filament is connected to a bridge circuit, and the device gives a readout in velocity, or a voltage output to a logger. Compensation for the air temperature is included. As there are no moving parts to cause friction, velocities as low as 0.4 m/s can be measured.

Pitot tubes

Pitot-static tubes comprise two tubes, one facing the air flow directly and the other perpendicular to the flow: they are normally contained in one measuring head (Figure 5.1). When air enters the tube facing the air stream, its kinetic energy is converted to pressure energy, therefore the pressure in the tube is equal to the static pressure in the duct plus the velocity pressure. The static pressure is measured by the tube perpendicular to the flow. The air is initially at velocity V which reduces to zero as it is brought to rest in the tube. The pressure differential ΔP is equal to the velocity pressure and is given by:

$$\Delta P = 0.5\rho_a V^2$$

where ρ_a is the density of the air (normally taken as 1.2kgm^{-3})

$$V = (1.66 \times \Delta P)^{0.5}$$

If ΔP is measured in a liquid manometer,

$$\Delta P = \rho l \times g \times h$$

where ρl is the density of the liquid in the manometer, h is the differential in the column heights, and g is the acceleration due to gravity (9.81ms^{-2}).

$$V = (16.35 \times \rho l \times h)^{0.5}$$

A probe may be inserted through a suitable hole to measure air velocity in a duct, and it is important that the size of the probe should be small relative to the duct, so as not to impede the flow significantly. Ideally there should be a section of straight duct before and after the pitot tube of ten diameters in length. As the air flow is not uniform across the duct, for best results a traverse should be made.

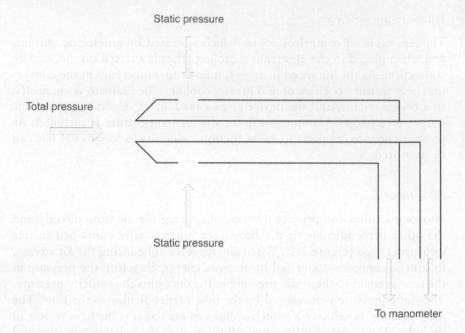

Figure 5.1 Pitot-static tube

Ventilation or infiltration rates

Air infiltration is the adventitious entry of outside air into a building, and may constitute a heat loss or a heat gain, depending on whether the ambient air temperature is higher or lower than that inside. Infiltration rates may be measured in air changes per hour (Ac/hr) and for well-sealed dwellings should be less than 1 Ac/hr. The heat loss (or gain) can be calculated from:

$$Q_{heat\ exchange} = 0.33 \times n \times V \times \Delta T$$

Where n is the ventilation rate in air changes/hour, V is the volume of the space and ΔT is the inside-outside temperature difference. It is therefore important to know the infiltration rate, which can be measured in two ways, using tracer gases or blower door tests.

Tracer gas method

This technique relies on the introduction into the room of a foreign gas which can be easily detected and whose concentration can be measured. Commonly used gases include CO_2 and N_2O, which are readily available in cylinders. The valve of the cylinder is opened to allow some of the gas to be injected into the room, and when sufficient gas is deemed to have entered, the valve is closed. The gas should be distributed uniformly throughout the space. After injection has ceased, the concentration of the trace gas will begin to fall as fresh air enters from outside, and some of the air containing the trace gas leaves through cracks in the fabric and window frames, etc. The rate at which the concentration falls is dependent on the rate at which the air in the room is changing, or in other words, the infiltration rate. The rate of change of concentration of the gas is a function of the amount of gas leaving the space, the amount entering, and the amount generated in the space.

$$V \times \frac{dC}{dt} = G + Q(C_e - C)$$

where
V = volume of room (m³)
G = generation rate of gas in room (m³s⁻¹)
C_e = external or background concentration of gas (kgm⁻³)
C = concentration of gas in room at time t (kgm⁻³)
Q = Quantity of outside air entering (m³s⁻¹).

When injection of gas into the space has stopped, $G = 0$. (Note that if we use CO_2, G may not be zero as some will be exhaled by the person carrying out the test if they remain in the room.)

$$V \times \frac{dC}{dt} = Q(C_e - C)$$

If the external gas concentration is zero, $C_e = 0$. (Note that C_e will not be zero if CO_2 is used.)

$$V \times \frac{dC}{dt} = Q \times C$$

Separating the variables:

$$\frac{dC}{C} = -\frac{Q}{V} \times dt$$

$$C_t - C_0 \times e^{-\left(\frac{Q}{V}\right)xt}$$

Taking logs:

$$\ln C_t - \ln C_0 = -\frac{Q}{V} \times t$$

where C_0 is the concentration at the beginning of the decay and C_t is the concentration at time t.

$Q/V = n$ = Number of air changes per unit time

Plotting $\ln C$ against t (in hours) gives a reasonable approximation to a straight line of slope $-Q/V$, which is the air change rate in air changes per hour.

Blower door test

To assess the infiltration and leak characteristics of a building, a blower door test can be carried out (see Figure 5.2). In the UK it is necessary to test a proportion of dwellings in any new housing development and all new non-domestic buildings in order to satisfy the Building Regulations, while it is not mandatory for most existing buildings. Full details are given in CIBSE Technical Manual 23 and the appropriate sections of the Building Regulations. The equipment consists of a false door which is positioned in a doorway and which contains a fan, flow-measuring apparatus and sensors to measure air pressures inside and outside the building. When the fan is switched on it blows air into the room and the air pressure inside will increase. Air will be forced out of the building through small cracks at the edges of doors, windows, floor–wall joints and so on. The greater the

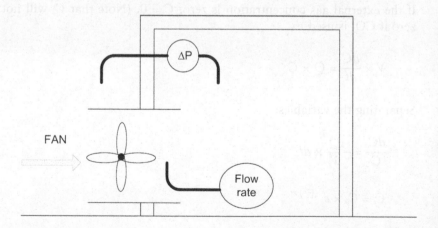

Figure 5.2 Blower door test

total air leakage through these cracks, the lower the excess pressure for a given flow rate. For each room there is a characteristic relationship between the flow rate, the pressure build up, and the leakage characteristics of the building. A variable-speed fan allows the flow rate and pressure to be measured over a range of flow rates. Pressure differentials of up to 50 Pa are normally used in order to obtain reasonable accuracy and flow rates. As these are far higher than encountered normally through wind pressure on the building, the results should normally be extrapolated to estimate the leakage rate at a more realistic pressure difference; alternatively, the raw leakage values measured at high-pressure difference can be used to make direct comparisons with other buildings. A very approximate assessment of the annual infiltration rate can be obtained by dividing the leakage rate at 50 Pa by 20. The relationship between pressure and flow is of the form:

$$Q = k(\Delta P)n$$

Where
Q = flow rate (m³s⁻¹)
ΔP = pressure difference (Pa)
k =constant unique to the room
n = constant unique to the room

$$\log_{10} Q = \log_{10} k + n \times \log_{10} \Delta P$$

Plotting $\log_{10}Q$ against $\log_{10}\Delta P$ should give a reasonable approximation to a straight line whose intercept is $\log_{10}k$ and slope n. Thus, values for k and n can be obtained and inserted in the equation to allow the flow rate to be estimated at a more reasonable pressure differential such as 5–10 Pa. It is usually advisable to test the building in both pressurising and depressurising modes (by reversing the fan), as the flow mechanisms may be different (for example, windows being pushed out against the frame). For comparison purposes the leakage rate is normalized in relation to the surface area S of the building to yield values in units of Q50/S (m³.h⁻¹).m⁻².

Water flow

A number of flow meter types are available to measure the flow of water and other fluids through pipes. They are generally based on an impeller-type mechanism whose speed of rotation is proportional to the fluid velocity. They are connected by gears to a dial on the outside of the pipe which can be read periodically, or connected to a pulse counter which sends an electrical signal to a logger and which is converted to flow rate.

Relative humidity

Individual measurements of relative humidity can be made using the traditional wet-and-dry bulb hygrometer, and tend not to vary greatly from place to place within a room. Electronic instruments using sensors employing the absorptive properties of lithium chloride and alumina are widely available. Handheld instruments or sensors that can be connected to data loggers are common, although regular calibration is required.

Heating efficiency

The heating energy delivered to the rooms through the central heating system (as opposed to the energy put into the heating system) can be measured using a heat meter (see Figure 5.3). A heat meter consists of: two temperature sensors, one measuring the temperature of the flow from the boiler, the other measuring the return temperature; a flow meter; and an integrator or data logger. It may be configured to give readings directly in kW and kWh, and may also feed data into a BMS. The heat delivered to the heating system in the building is defined by

$$Q = \dot{m} \times C \times \Delta T$$

Where
Q = the heat delivery rate (W)
$\dot{m}$ = the mass flow rate of water (kgs^{-1})
C = the specific heat of water (Jkg^{-1}K^{-1})
ΔT = the difference between the flow and return temperatures (K).

The quantities measured are $\dot{m}$ and ΔT, while C is a constant (approx. 4,200 JKg^{-1}K^{-1}).

If the rate of fuel consumption is recorded simultaneously, the efficiency of the heating system can be estimated as described below.

Example of heat meter calculation

When used for an extended period the heat meter can yield useful information about the building. If a single heating system covers the entire building, and the fuel flow to the boiler is metered, then the heat meter can be used to assess the efficiency of the boiler and heating system. Assume the system is operating steadily for a few hours, and that the following data is collected:

- average temperature of water from boiler to system   82 °C
- average return temperature to boiler   75 °C

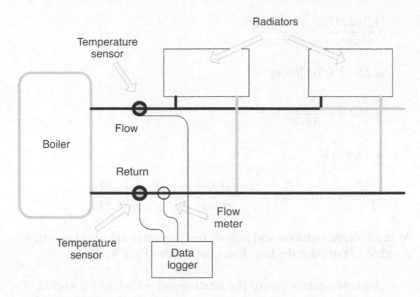

Figure 5.3 Schematic of a heat meter set-up to measure the heat put out by a central heating system

- average water flow rate in heating system 1.6 kgs^{-1}
- average outdoor air temperature $7\,^\circ\text{C}$
- average Indoor air temperature $23\,^\circ\text{C}$
- gas consumption over the measuring period (6 hours) 35.5 m^3
- specific heat of water $4190 \text{ Jkg}^{-1}\text{K}^{-1}$
- calorific value of gas 38.9 MJm^{-3}
- cost of gas 4.5 pkWh^{-1}

The efficiency of the boiler is equivalent to the heat output/heat input.

> *Heat output* = $(m \times \dot{C} \times \Delta T)$
> = $1.6 \times 4{,}190 \times (82 - 75) = 46{,}928 \text{W}$

The heat input is based on the gas input multiplied by the calorific value.

> *Heat input over 6 hours* = $35.5 \times 38.9 = 1{,}380.9 MJ$

The input and output should be converted to the same units; in this case W are chosen.

The rate of heat input is 1380.9 MJ in 6 hours, which must be converted to J/s.

$$\frac{1,562MJ}{6hours} = \frac{260.3MJ}{hour}$$

$$= 260.3 \times 10^6 J/hour$$

$$= 260.3 \times \frac{10^6}{3600} J/s$$

$$= 63,933\,W$$

$$Efficiency\ of\ boiler = \frac{useful\ energy\ out}{energy\ in} = \frac{46,928}{63,933} = 73.4\%$$

As the average outdoor and indoor temperatures are also known, it is possible to estimate the heat loss coefficient of the building.

Average heat output of the heating system (above) = 46,928 W

Average indoor-outdoor temperature difference = 23 − 7 = 16K

$$Total\ heat\ loss\ coefficient\ (TLC) = \frac{(heat\ loss)}{(temperature\ difference)}$$

$$= \frac{46,928}{16} = 2,933\,WK^{-1}$$

Knowing this, it is possible to use the degree-day method to estimate annual energy consumption, as shown in Appendix 1. The method is approximate and does not take into account thermal storage in the building fabric or variation in internal heat gains, but in the absence of other methods it can give a reasonable estimate of the thermal performance of the building.

U-value measurement

U-values will normally be read or deduced from the building documentation, if available, but in some cases it may be desirable to measure them – where there is uncertainty as to whether insulation has been added to a wall, for example. The thermal transmittance of a building element (U-value) is defined as the 'average heat-flow rate per area in the steady state divided by the temperature difference between the surroundings on each side of a system' (ISO 7345). The units are $Wm^{-2}K^{-1}$. Accurate methods for measuring U-value use a guarded hot-box technique and are described in BS EN ISO 8990, but an approximate value can be obtained as shown in Figure 5.4 by logging the temperatures either side of the wall and the heat flow through it. The apparatus required includes a data logger, temperature sensors such as thermocouples or thermistors, and heat-flow sensors. A heat-flow sensor comprises a thin disc of material with a constant thermal resistance which has embedded in it a series of thermocouples to measure the temperature difference through the disc. A millivolt output is produced, which is proportional to the heat flux. The heat flux and temperature difference should be logged at intervals of 10–15 minutes over a few days to minimize the effects of thermal storage. The average temperature and heat-flow values over the period can be used to calculate the U-value, which is given by:

$$\frac{Q}{A} = U \times \Delta T$$

$$U = \frac{Q}{A \times \Delta T}$$

Where
Q/A = the heat flux measured by the sensors (Wm^{-2})
ΔT is the temperature difference (K).

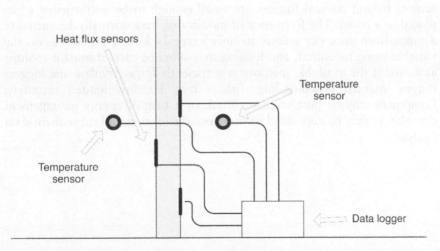

Figure 5.4 Measuring the U-value of a wall: the temperature and heat flux sensors

There are a number of potential sources of error in such measurements:

- heat flow may not be one-dimensional, particularly near corners
- thermal storage effects in the wall may lead to erroneous readings
- the section of wall chosen for the measurement may not be representative of the whole: for example insulation in the wall may be damaged or may have settled out.

They can be reduced by:

- positioning the heat flux sensors well away from potential disturbances
- monitoring over long periods to reduce storage effects
- making measurements in several places
- using several heat flux sensors.

Data loggers

A data logger is a piece of electronic equipment that has a series of channels, which can be configured to accept inputs such voltage, current, pulses, and digital inputs. It has a memory for storing the data collected, and facilities to download the data to a computer or printer, possibly through a Bluetooth connection. Once downloaded, the data can be processed using a spreadsheet or proprietary software. A large range of models is currently available, and while some are designed for specific applications such as collecting electricity consumption data from a small number of input channels, others are more versatile and can accept various forms of data into as many as 60 input channels. Some specialized temperature and humidity sensors having integral loggers are small enough to be unobtrusive when placed in a room. The frequency of monitoring may normally be varied in a range from once per second to once every 24 hours, depending on the variable being measured, and logging may often be carried out for months at a time if the available memory is sufficiently large. While some loggers require manual downloading, others may be downloaded remotely. Equipment may be purchased or hired, or a firm of energy management consultants may be appointed to carry out the logging and subsequent data analysis.

6   Organization and implementation

An organization may simply commission an energy audit, act on some of its findings, and forget about energy management for the next few years. While this approach may achieve short-term savings, for lasting and increasing savings a strategic approach is required, and in today's energy markets a long-term view is essential. For long-term effectiveness, energy management should be thought of as a continuing process, not as a single activity to be carried out once and then forgotten. With this in mind, in 2015 the UK government set up ESOS (Energy Savings Opportunity Scheme), which is described in Chapter 7.

For action to be effective, commitment is required from all levels in an organization. In this respect it is important to develop an action plan which helps to establish priorities for improving how energy is managed in the operation. Central to the action plan is a need to assess the organization's position regarding energy management, and an assessment of current operating practices is therefore essential. The level of commitment to controlling energy can be assessed and appropriate action taken in areas where there are deficiencies.

The organization of energy management activities may take various forms, depending on the size of the enterprise, the amount spent on energy, and the corporate structure. An in-house energy manager may be a specially appointed person whose sole responsibility is energy, or, as happens quite often, a member of staff who has other main duties besides energy. Alternatively, outside consultants may be employed. Many energy management companies exist, offering a range of services including carrying out energy audits, running utility services such as combined heat and power (CHP) plants, financing investments in energy plant, or providing heat and electrical power. Forms of contract include fixed-fee arrangements (e.g. an energy audit carried out for an agreed fee) and those where the consultant takes as their fee an agreed proportion of any savings, often on a no-savings, no-fee basis. Since many energy saving measures tend to be long-term in nature with a number of years to payback, the latter tend to be long-term contracts of from five to fifteen years. Whether the energy manager is based in-house or is an outside consultant, they need to:

- develop an energy audit system and carry out energy audits
- devise tough but realistic targets for energy savings
- provide technical advice
- be able to assess the cost-effectiveness of energy-saving measures
- monitor developments in energy conservation
- advise on government funding such as grants
- keep abreast of political, legislative and regulatory measures affecting energy use and costs.

Energy management can be a 'tough sell', especially in times when energy prices are low, so besides technical and commercial knowledge of buildings and energy supply, an energy manager also needs good interpersonal skills in order to be able to motivate management and other staff to invest in and implement energy-saving measures.

Functions of the energy manager

The functions of the energy manager comprise the following actions:

- assess the level of awareness and commitment from management
- get commitment to energy management from the top of the organization
- identify the corporate structure and management style
- assess the level of awareness within the company, and raise it if necessary
- motivate others in the organization to help improve energy performance,
- devise an energy policy
- assess current performance levels
- set up short, medium and long-term objectives and develop procedures to accomplish these objectives
- set up a system to measure and document energy use, and forecast future consumption
- agree targets for improvement, along with the associated budgets
- set up a programme to review objectives and achievements at regular intervals.

Assessing the level of awareness and commitment from management

It is important to know where the organization stands from the outset, and this can be ascertained by asking a number of key questions:

- is there an energy policy
- how aware of energy use are the workers and managers
- how are energy improvements financed, is there a budget or do funds have to be fought for

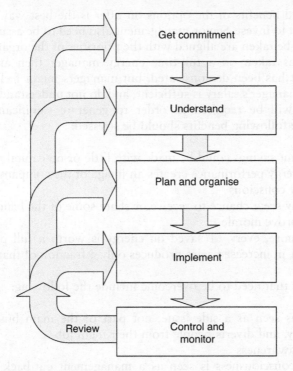

Figure 6.1 Organizing energy management

- are there accurate records of energy use
- have any energy management activities been carried out previously
- are there any plans for improving efficiency
- what are the opportunities for intervention to improve efficiency
- what are the risks.

Getting commitment

Most organizations with core activities such as financial services, manufacturing, software design or local authority services tend to regard energy as an unwelcome but unavoidable cost to the business, and have no particular interest in it. However, price rises over recent decades have highlighted the true cost of energy, and it can no longer be ignored. Schemes such as ESOS (See Chapter 7) are mandatory and organisations that do not comply will be fined, but at present ESOS only requires energy audits to be carried out, it does not force companies to act on the findings. In difficult economic times, obtaining commitment to significant investment in energy efficiency is never going to be easy, and a careful and detailed economic analysis of the costs and benefits of the options on offer is the best way to convince management to invest. Senior management also need to be assured that any measures to be taken are aligned with the priorities of the organisation. If a company has taken on a full-time energy manager then at least some commitment has been demonstrated, but managers might feel that paying the energy manager's salary is sufficient, and do not understand that further investment will be required in order to generate significant long-term savings. The following benefits should be stressed:

- significant savings could be made with little or no capital expenditure
- good energy performance creates an image of the company as environmentally conscious
- staff may see a chance to receive or share some of the benefits and this will improve morale
- furthermore, every £1 saved on energy is worth a full pound, while every £1 in increased sales produces only a fraction of that in profit.

The barriers that need to be overcome include the following:

- energy is seen as a side issue, not part of the main business of the company, and diverts people from their main job
- lack of awareness
- energy consciousness is seen as a management cut-back and penny-pinching.

A further obstacle in the commercial sector is that only 10 per cent of office buildings are owner-occupied, which acts as a serious deterrent to investment in energy efficiency. Landlords may be unwilling to invest in measures which reduce their tenant's energy bills, while short-term tenants have no incentive to invest in measures that might only pay off after they have left. Money to invest in energy-saving measures is not usually handed over without careful scrutiny of the costs and benefits, using some of the techniques discussed later in this chapter. Investment in energy efficiency is often given low priority, as it is usually identified with savings rather than investment. In any organization there will normally be a number of projects competing for limited funds, and investors will quite naturally wish to make the best use of their money. Financial controllers will be looking to obtain the optimum benefits from each investment with the minimum risk, and it will be essential for the energy manager to understand how the enterprise allocates money between capital and revenue budgets. Also, it may be difficult to identify an appropriate budget under which energy-saving schemes could come. Capital plant, development or maintenance budgets may be suitable but the budget holders may have other priorities.

Opportunities

While there may be a small ongoing energy budget, other events may present opportunities to invest in energy-saving measures. These include:

- single actions identified at particular times, e.g. refurbishment of a building, particularly when it involves a change of use; this may include opportunities to install CHP, upgrade the insulation of the building, install double glazing, or use natural ventilation in place of air conditioning
- programmes carried throughout an organization, such as a hotel chain upgrading the lighting in all its hotels. It may also be possible to hang other measures on the general programme, such as installing sub-metering or better controls.

Risks

There is usually some element of risk involved, but for most measures this will not be a technical risk but a financial risk – the extent to which the project is exposed to variations in factors which affect cash flows such as energy tariffs and interest rates. While energy costs follow a generally upward trend, there can be significant short-term variations, and committing to purchase large quantities of energy at a particular time can sometimes lead to unwelcome losses.

Identifying the corporate structure and management style

The way in which the organization operates will dictate the approach that has to be made to motivating people and obtaining appropriate budgets. The structure and management style may range from very hierarchical to strictly egalitarian. It is also important to identify those responsible for energy-consuming plant. Where this includes jointly used equipment such as photocopiers, it can be difficult to track down the responsible individual; it may be the maintenance/engineering department, the building manager, a facilities manager or in fact no one in particular. The implications are not trivial: a photocopier left on all night will use enough power to produce about 1,500 copies. Consider the annual cost of this, and compare it with the cost involved in simply allocating someone to switch it off before the office is closed.

Assessing and raising staff awareness

Raising general staff awareness of energy issues is an important element of obtaining commitment. Energy is in the news a great deal these days, but staff still need information on how energy use affects them in their organization and their particular job. A general 'save energy' approach does not necessarily help staff identify the areas where they can best help. A start can be made using a home energy assessment survey, to identify how aware staff are of energy savings that can be made in their own homes. Many of the lessons learned can then be applied to the workplace. Methods of raising awareness can take a number of forms, including presentations, training sessions, use of company newsletters, email and websites, suggestions schemes, poster campaigns, and 'save it' stickers. The ultimate aim is to integrate energy efficiency into everyday working practices and make energy-saving activities such as switching off lights part of the everyday routine.

Motivating others

Raising awareness is part of motivation but motivation also includes other aspects such as personal, departmental or team reward schemes for improved performance, linked to monitoring and targeting activities. Motivating management may require demonstrations of potential energy savings using detailed costed examples, as budgets will need to be agreed. People will generally not change to work in an energy-efficient way for its own sake – they need to see that there is some benefit to themselves for doing so. Such benefits could include greater profitability and competitiveness for the company as a result of lowered costs, hence more job security. Benefits to individual departments, such as increased budgets or a chance to utilize a share of the energy savings, may also act as an incentive. Schemes such as 'energy saver of the month' and suggestions schemes have also been shown to be effective. An important point here, as with raising awareness, is that a one-off approach tends to fall off in effectiveness after some time, and new approaches or initiatives need to be introduced from time to time in order to maintain the momentum. Communicating the results of any initiatives back to those involved is also crucial in maintaining interest. Articles in a company newsletter or on the website can highlight improved performances of groups and individuals.

Devising an energy policy

A company's energy policy needs to be agreed with senior management, since its implementation will involve important investment decisions. The objectives of the energy policy will depend to some extent on the current attitudes towards energy, the organization's business priorities and its existing energy performance. The following steps may be taken.

Assessing current level of performance

It is essential to know how well the organization is succeeding at present in the management of energy:

- Are space temperatures appropriate, too hot or too cold?
- Is heating left on when not required, e.g. in the evening and at weekends?
- Are lights left on unnecessarily?
- How well are the building fabric and services plant maintained?
- Is small electrical equipment, such as computers, routinely left on overnight?

The answers to these questions will point the way towards short-term improvements that will help to inspire confidence in the energy management process.

Setting up short-, medium- and long-term objectives

It is important to set up specific objectives rather than vague commitments to 'cut energy' or 'be more efficient'. These objectives will vary depending on the time scale being considered. A simple low-cost change that gives a quick result will be very effective in developing confidence in the energy management process, and it may be possible to use the savings from it to finance longer-term measures.

Examples of short-, medium- and long-term objectives

Short term:

- mend faulty items of energy-using plant,
- ensure items of equipment are switched off when not in use or at the end of the working day. Simple time switches to turn items off at the end of office hours will pay for themselves in weeks,
- ensure rooms are not heated or cooled excessively,
- sticker campaign for light switches,
- report back on improved performance.

Medium term:

- set a target of 10 per cent energy reduction in a year,
- instigate a better maintenance regime,
- report on costs and benefits of sub-metering,
- appoint local 'energy champions'.

Long term:

- set up a five-year energy reduction target of say 20 per cent,
- set up monitoring and targeting software for cumulative sum (CUSUM) analysis and similar activities,
- report on feasibility of replacing boilers with CHP,
- provide professional training for energy champions.

Setting up a system to measure and document energy use

Energy consumption information must be authoritative if it is to be used to inform the investment policy: that is, it must be obtained from reliable sources, and be accurately documented. Such sources include detailed utility bills, half-hour metering information, and logged data from the building management system (BMS). Future consumption can be forecast using such techniques as the CUSUM analysis described later in this chapter.

Agreeing targets for improvement, along with the associated budgets

The measures that require to be implemented in order to meet the energy reduction targets should be properly costed and the benefits presented realistically and accurately, otherwise management is unlikely to agree to release funds. Targets which are imposed, not agreed, will have little effect since no one will feel committed to honour them. If possible, a range of options should be presented.

Setting up a programme to review objectives and achievements at regular intervals

The appropriate timescale for review should be selected. Too short a time scale will not allow any savings to be made or demonstrated, while if too long a period is chosen, people may lose interest in the interim. If new plant is installed, such as a condensing boiler, it will be some months before the benefits are evident in the energy bills. With other measures it is likely to take a matter of years for the benefits to materialize, and it is pointless to review the objectives before the measures have had time to take effect. The plan should include a regular commitment to review progress, as needs may change over time, for example due to changes in use of parts of the building, changes in the level of staffing, or an increase or decrease in production levels. It is important that regular reviews are fed back into the entire energy management process in order to take into account changes in energy supply, legislation, changes within the company, and financial arrangements. A large amount of free literature on this subject is available from the Carbon Trust (www.carbontrust.co.uk). Working to an agreed standard is a useful way of achieving and demonstrating commitment. BS16001:2009 shows in detail the requirements to be met in order to achieve the British Standard for Energy Management, and enlarges on the outline of organized energy management shown here.

Monitoring and targeting

In order to provide quantified data to management to inform investment decisions concerning energy, the energy use needs to be monitored, and methods developed to devise realistic targets for energy savings. These come under the general heading of monitoring and targeting (M&T). They form essential elements in understanding and controlling energy use, and also provide information to feed back to management on the performance of a building. The purpose of M&T is to relate energy consumption to some variable such as the weather or production output, in order to understand better how energy is used and help identify avoidable waste. Data collection is an essential element of M&T, and may be carried out automatically or manually, depending on the kind of data being collected, the technology available, and the budget. Analysis of the data will suggest clear lines of investigation for producing savings, which can easily be quantified. It is important to set targets that are realistically achievable, but not so low as to be meaningless. An overall reduction of 5 per cent in energy costs over five years is unlikely to justify the time and money put into it, while a target of 30 per cent, although extremely tough if no special measures are taken, may be realistic if a budget for significant new plant such as CHP is available. When an organization is just beginning an energy management programme, targets of 5–10 per cent may often be achievable at low or zero cost by simple common-sense measures such as switching off unwanted items of plant, but these are one-off items that are unlikely to provide further opportunities. Targets should be discussed and agreed between the relevant parties, and based on sensible calculations of performance. The smaller the unit of consumption being targeted the better, as greater control can be exercised, and frequent reviewing of performance enables problems to be identified quickly. Examples of such problems include faulty timers which allow heating to remain on 24/7 even over Christmas holidays, badly calibrated temperature sensors, or broken control valves.

A key element of M&T is accurate forecasting of the expected energy consumption; exceptions can then be reported and acted upon. A simple way of reporting exceptions is to use an overspend league table, which shows the overspend on specific items of energy consumption over a fixed period, in descending order of cost. High-overspend items can be discussed with those responsible, and further analysis carried out, or if the reason for the overspend is obvious, any necessary remedial action can be taken. A week is often a suitable reporting interval, but daily or monthly reporting may also be used, depending on the circumstances. Calculations of expected consumption are based either on precedent (direct comparison with previous periods) or activity: in other words, related to the driving factors, such as weather or production quantity. Precedent-based targeting is usually based on monthly figures and year-on-year comparisons, and suffers from the weakness that significant changes in the weather may occur from one year to the next, which are not taken into account in this method. It also does not take into account changes in working practices such as variations in opening hours and weekend working. This makes the determination of 'exceptions' rather difficult. Activity-based targeting enables changes in the weather to be allowed for, and as we now have relatively easy access to weather data from a large number of locations, it can be used to provide more incisive analysis of the performance of a building. Heating and cooling are among the greatest causes of energy use in buildings, and are related directly (but not exclusively) to the weather. Plotting a graph of heating energy use and degree days as shown in Chapter 3 is one of the simplest but most instructive ways of analysing performance, and M&T software can be configured to produce such plots automatically. For plots of energy use against degree days the relationships are linear, of the general form $y = mx + c$, where m is the slope and c is the intercept on the y axis. Similar plots can be drawn for process energy, such as energy use against throughput for an oven or drying plant, but they are not invariably linear. Process energy plots may in fact reveal that consumption is unrelated to throughput, but this knowledge in itself is useful as it may point to a better form of operation or control in order to save energy. For batch processes, plant is often left on between batches, which could lead to large amounts of energy being wasted. When plotting energy use against temperature the user normally has the option to set the intercept at zero. For heating degree days this will not usually be the case, and the intercept will depend on a number of factors, including the base temperature used for the degree-day calculation. For processes there may be a fixed amount of energy use plus a throughput-related element (c and m respectively in the $y = mx + c$ plot). It is also essential to be able to measure the extent to which the target has been approached. This is done by analysing utility bills, but may also be done by monitoring, using some of the instruments and techniques described in Chapter 5. Smart meters which provide this kind of data for the customer and also to the energy supplier are now becoming more common, and will be universal for domestic properties in the UK by 2020. They may also be linked in with automatic meter reading (AMR) systems which have been trialled in a number of places. An essential feature of both these is that half-hour meter readings can be made, giving both the supplier and the user much finer-grained data than was previously available. For large users half-hour meter reading is essential since the cost per unit varies with time of delivery; such a detailed knowledge of the consumption pattern may give the user the opportunity to shift loads to a time of day when energy is cheaper. It may also point the way to avoiding or minimizing maximum demand charges by shifting some loads to times of lower demand. A conventional meter costs about £500 including installation, and if a payback of two years is expected, then each meter needs to save £250 per year. A minimum of 5 per cent savings should be aimed for, so that the utility cost through each meter installed should be at least £5,000 to make it worthwhile. Analysis of the data using CUSUM or other methods may be done using proprietary software (some of which is available as an add-on to the BMS) or using a bespoke spreadsheet developed by the energy manager.

CUSUM analysis

CUSUM stands for the cumulative sum of deviation: in other words, deviation from the expected consumption. CUSUM analysis is a tool of M&T, and in its simplest form the cumulative sum of energy consumption is plotted against time. The data required can be obtained from energy bills, from BMS data, or from a monitoring exercise. The slope of the curve will change from time to time, depending on the conditions. For example, gas consumption may cover heating, cooking, and domestic hot water; in summer there will be no heating, so the consumption will be much reduced and the slope will be lower (see Figure 6.2).

The slope effectively gives the rate of energy use. Since there is no space heating in summer the slope reduces considerably during those months, and is approximately constant over that period, indicating that the cooking energy is roughly constant. Ideally, data from more than one year should be available, so that a straight comparison from one year to another can be carried out, allowing the seasonal changes in consumption to be observed (see Table 6.1). The results are best shown in histogram form (as in Figure 6.3). Here, only the raw heating energy data is used and differences in the weather over the same period are not taken into consideration. A simple way to take account of the weather is to calculate the energy consumption per degree day (see Chapter 3), so that kWh/degree day can be used and a direct and meaningful comparison can be made (see Table 6.2 and Figure 6.4).

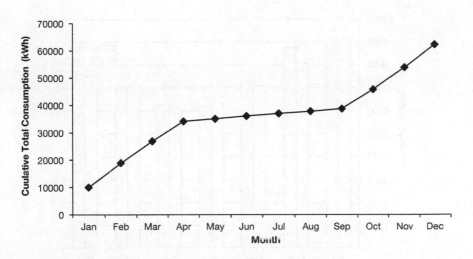

Figure 6.2 Cumulative energy consumption (total of space and water heating and cooking)

Table 6.1 Monthly heating gas consumption and degree days

Month	Previous year heating energy consumption (kWh)	Previous year degree days	This year heating energy consumption (kWh)	This year degree days
Jan	9100	598	9069	576
Feb	8765	488	8054	453
Mar	7250	354	7073	324
Apr	6200	269	6329	272
Oct	6243	333	6154	327
Nov	6989	442	7078	434
Dec	7600	528	7509	524
Annual total	52147	3012	51266	2910

Table 6.2 Energy consumption in kWh/DD for current and previous year, deviation and CUSUM

Month	kWh/DD This year	Last year	Deviation	CUSUM
Jan	15.74	15.22	0.52	0.52
Feb	17.78	17.96	−0.18	0.34
Mar	21.83	20.48	1.35	1.69
Apr	23.26	23.05	0.21	1.9
Oct	18.82	18.75	0.07	1.97
Nov	16.31	15.81	0.5	2.47
Dec	14.33	14.39	−0.06	2.41

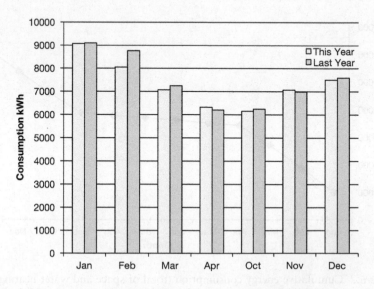

Figure 6.3 Heating energy consumption for two years compared

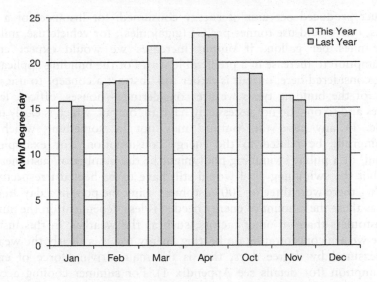

Figure 6.4 Energy consumption per degree day – comparison of two years

CUSUM analysis is relatively simple to perform, as it can be carried out using spreadsheets or standard M&T software. It can be used in three main ways:

- to set realistic targets for energy consumption
- to assist in the diagnosis of persistently excessive consumption
- to quantify cumulative savings.

The analysis assumes that if the building is operated correctly, energy consumption will be predictable. The calculation is based on energy consumption over a specified measuring interval: monthly, weekly, daily, or even hourly intervals can be used, depending on the requirements and the data available. Whichever interval is selected, at the end of each period the actual quantity of energy used must be measured and an estimate made of the amount of energy that should have been used: in other words, the expected consumption. The deviance for the period is obtained by subtracting the expected consumption from the actual consumption. Adding the deviance for the last period to the running total of deviance gives the CUSUM, or cumulative sum of deviation. The example on the next page shows the operation of CUSUM analysis in the context of heating a building. The actual consumption data can be obtained from meter readings, but it is more difficult to calculate the expected consumption as the drivers of energy consumption need to be known. In the case of manufacturing industry a useful measure of expected consumption is the amount of product produced per unit of energy consumed: for instance for a steel works, one would use tonnes per GJ (gigajoules); for vehicle use, miles or tonne-miles per gallon. If output increases we would expect energy consumption to increase in a predictable way. For the building applications being considered here, output is rather too abstract a concept to use, since many of the building types we are considering – houses, offices, leisure centres and so on – do not necessarily have an 'output' which is easily measurable. In any case, the 'output' may not be something which can meaningfully be related to the energy consumption. For example the 'output' of a public swimming pool might be the number of customers per day, but the swimming pool would still have to be heated irrespective of whether there were three or 300 customers using the pool in a day. In cases such as these the amount of energy needed is less dependent on the number of customers than on other factors, such as the weather. In this instance where we are concentrating on heating energy use it is clearly the weather, as measured by degree days, that is the main driving force of energy consumption (for details see Appendix 1). For summer cooling a corresponding measure, the cooling degree day, can be used. In a well-regulated building the heating energy consumption per degree day should be roughly constant. Figure 6.5 shows the data for one year from Table 6.2 redrawn in this way. Since the heating energy consumption is zero over the non-heating season, points from this period are omitted. (Energy consumption divided by zero degree days would give infinity.)

This form of plot already yields additional information. It may be observed that the energy consumption per degree day increases from winter

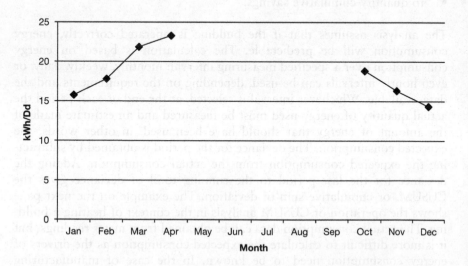

Figure 6.5 Heating energy consumption per degree day, plotted by month

to spring, and decreases from autumn to winter. This may occur for a number of reasons. One is that in winter it is likely that the load factor will be greater and the boilers will be running more efficiently; another possibility is that in the coldest weather the occupants are more careful about closing doors and windows. For periods where the consumption is less than predicted, it is worth investigating what happened, since good practice carried out in those periods should be repeated if possible.

In order to pursue the CUSUM analysis the expected energy consumption must be calculated. One approach is to take the average consumption per degree day for the previous year, calculated here as 17.3 kWh/DD. Using this figure, the expected energy consumption for the corresponding period in the following year (and the deviation) can be calculated. These values are shown in Table 6.2 and Figure 6.6. The CUSUM analysis is shown in graphical form in Figure 6.4 where it may be observed that for most months the deviation is increasing, demonstrating that consumption is worse than the previous year. During two months the deviation decreases, illustrating better performance than previously, but the overall trend is upwards away from zero. A sharp increase in deviation between February and March indicates that something serious has occurred – a broken sensor or valve, for example.

During the summer months the deviation falls considerably, suggesting that the main problems lie with the space heating. From November to December the deviation falls, possibly following some remedial action to improve control. The cumulative deviation at the end of the year is significant, showing that there has been an overall deterioration in performance

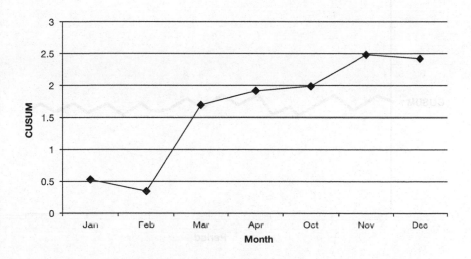

Figure 6.6 CUSUM analysis for the example given

since the previous year. The reasons for this may be manifold, and detailed study of the CUSUM analysis is of help in finding them. In a well-controlled building the 'expected' consumption for this year should be roughly the same as the previous year, after allowing for degree day differences. Although inevitably there will be slight positive and negative deviations, the expected consumption on average will follow a line parallel with the x-axis. If the CUSUM from the current performance characteristic is correctly derived and the building is well-regulated, the CUSUM chart will be similar to that in Figure 6.7. There are slight variations from the norm from month to month, but on average the trace runs level. If the characteristic has been set too leniently, there will be a general downsloping tendency (Figure 6.8). On the other hand, if the characteristic is set too harshly there will be a persistent rise in the graph (Figure 6.9). The shape of the chart gives an indication of how well the building is performing. An upward bend in the CUSUM chart indicates the onset of waste, and a downturn shows that savings are being made. The savings to date can be read off by measuring the vertical drop of the CUSUM plot (Figure 6.10). This figure also demonstrates a significantly improved performance from halfway through the period, as a result of changing the heating system controls. Normally after such a period of sustained improvement the characteristic would be reset, effectively creating new targets. This will occur naturally if a rolling table is used rather than annual tables with year-end breaks.

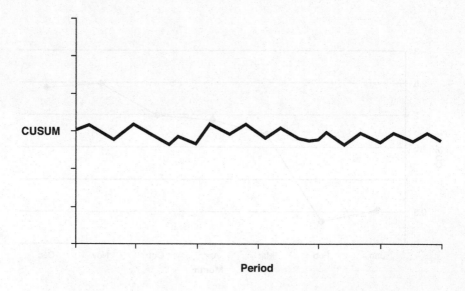

Figure 6.7 CUSUM of well-regulated building

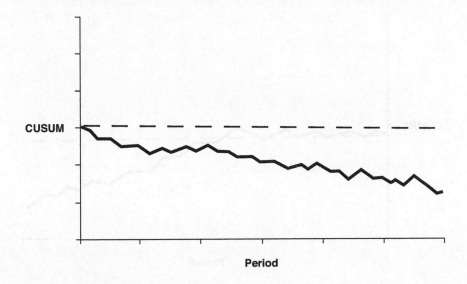

Figure 6.8 CUSUM of characteristic that is too lenient

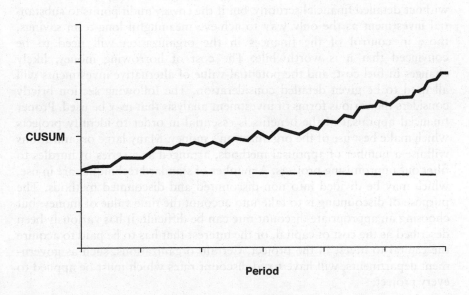

Figure 6.9 CUSUM where the characteristic is too harsh

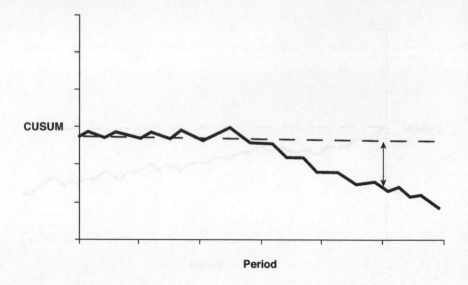

Figure 6.10 CUSUM showing cumulative savings (arrow)

Financial appraisal

Relatively small investments in energy management may be authorized without detailed financial scrutiny, but if the energy audit points to substantial investment as the only way to achieve meaningful long-term savings, those in control of the finances in the organization will need to be convinced that it is worthwhile. The cost of borrowing money, likely changes in fuel cost, and the potential value of alternative investments will all need to be given detailed consideration. The following section briefly considers the various forms of investment analysis that may be used. Proper financial appraisal of the benefits is essential in order to identify projects which make best use of the organization's money. Many large organizations will use a number of appraisal methods, arranged as a series of hurdles to filter out unpromising projects. A number of standard techniques are in use, which may be divided into non-discounted and discounted methods. The purpose of discounting is to take into account the time value of money, but choosing an appropriate discount rate can be difficult. It has variously been described as the cost of capital, or the interest that has to be paid to acquire the capital to invest in the project. Certain organizations, such as government departments, will have fixed discount rates which must be applied to every project.

The methods examined here include:

* payback period
* gross return on capital
* net return on capital
* gross average rate of return
* net average rate of return
* net present value.

Payback period

Payback period is the simplest of all to understand and to calculate. The capital cost of the project is simply divided by the expected annual savings to give a value in years. The advantages are that it is easily understandable, and the calculation of payback period is simple.

The disadvantages are that it does not take into account the timing of costs and benefits, likely residual value of assets at the end of the lifetime of the project, or savings accruing after the payback period (see Table 6.3 and Figure 6.11).

Example of the payback method

Table 6.3 Simple payback method calculation for two energy-saving measures

Year	Measure A Cash out	Cash in	Cum. net cash in	Measure B Cash out	Cash in	Cum. net cash in
1	1000	0	−1000	1500	0	−1500
2	0	200	−800	0	300	−1200
3	0	200	−600	0	300	−900
4	0	200	−400	0	300	−600
5	0	200	−200	0	300	−300
6	0	200	0	0	300	0
7	0	200	200	0	300	300
8	0	200	400	0	300	600

Measure A requires investment of £1,000 and gives savings of £200 per year. Measure B requires £1,500 investment, giving savings of £300 per year. While both measures have the same payback period, with the larger investment the cumulative savings after the payback period are much higher, therefore on this simple basis it would be considered more worthwhile. The cost of paying back money for the investment over a number of years, possibly with varying interest rates, is not taken into account here.

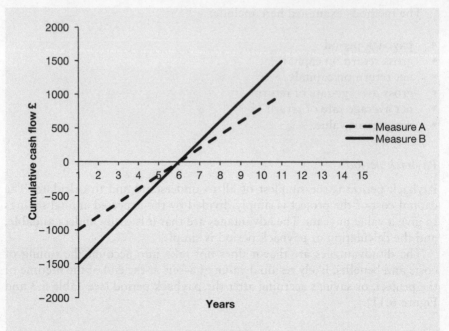

Figure 6.11 Cumulative non-discounted cash flow of two energy-saving investment measures with the same payback period

Accounting rate of return

This is also known as the average annual rate of return on investment, and concentrates on profitability. It considers earnings over the entire life span of the project but does not take into account the timing of those earnings. The two basic variants to this method are:

- average gross annual rate of return
- average net annual rate of return.

The average gross annual rate of return is defined as the average proceeds per year over the life of the assets expressed as a percentage of the original capital cost. The average net annual rate of return is defined as the average proceeds per year, after allowing for depreciation over the life of the assets, expressed as a percentage of the average value of the capital employed. The advantages of these methods are that:

- they are easy to understand and compute,
- they emphasize profitability, as returns over the whole life of the assets are taken into account.

The concept of return on capital employed is often considered the most important yardstick used in the measurement of business performance. The disadvantage is that no acknowledgement is made of the timing of costs or receipts, and irregularities in the cash flow are smoothed out by averaging. Once averaged, no indication is given of the time span of the return.

Examples of rate of return calculations

1 Gross return

Capital cost of project	£120,000
Total net cash inflow over 5 years	£170,000
Average net cash inflow per annum	£34,000
Average gross return (capital cost £120,000)	£34,000/£120,000
Average gross return	28.3%

2 Net return

Capital cost of project	£120,000
Residual value of assets at end (of 5 years)	£50,000
Total net cash inflow (over 5 years)	£170,000
Less depreciation	(£70,000)
Net return	£100,000
Average net cash inflow per annum (5 years)	£20,000
Average net return on average capital employed (£70,000)	£20,000/£70,000
Average net return	28.5%

Discounted cash flow (DCF)

DCF methods of investment appraisal involve discounting future outflow and inflows of cash back to present day values, thus establishing a common base for the comparison of investment alternatives. The difference between these and the methods already described is that they acknowledge the importance of timing, so that funds invested in the future have less impact than funds invested now, and that funds received (or savings made) early on in the lifetime of a project are worth more than funds received or savings made later. There are two basic DCF methods:

- net present value
- internal rate of return.

The present value of the funds invested is compared with the present value of the net cash flows expected to be generated over the life of the investment. An example is shown in Table 6.4. The following information is required:

- initial cost of the project
- cost of supplying the capital required, that is, the minimum rate of return required from the investment
- the values and timings of future cash flows for the total expected life of the project
- reference table of discount factors.

The method of computation is as follows:

1   Select the appropriate discount factor.
2   Calculate the present day value of each year's cash inflow by multiplying the values of those inflows by the appropriate discount factors.
3   Sum the values calculated to give the total present day value of future cash benefits.
4   Calculate the present day values of future cash outflows and add the sum of these to the initial cost of the project.
5   If the total arrived at under 3, the total present day value of all future returns, exceeds the total arrived at in point 4, the present day value of the investment costs, the project should be accepted; if not, it should not be pursued. The method is particularly useful where several alternative projects are being considered, as the one showing the highest positive result is the best performer and is likely to be the one accepted.

Example of a discounted cash flow calculation

Three projects all have the same initial cost, and the same total cash flow in: the difference is in the timing of the cash flows. In A the greatest inflows are at the beginning, in B at the end, while in C the cash flow inwards is the same each year.

Table 6.4 Discounted cash flow example

Project A			
Year	Net cash flow	Discount factor @10%	Present day value
0	−100000	1.0	(100000)
1	50000	0.909	45455
2	45000	0.8264	37188
3	35000	0.7513	26295
4	30000	0.6830	20490
5	20000	0.6209	12418
Total cash flow	80000	Total NPV	41846

Project B

Year	Net cash flow	Discount factor @10%	Present day value
0	−100000	1.0	(100000)
1	10000	0.909	9091
2	15000	0.8264	12396
3	35000	0.7513	26295
4	50000	0.6830	34150
5	70000	0.6209	43463
Total cash flow	80000	Total NPV	25395

Project C

Year	Net cash flow	Discount factor @10%	Present day value
0	−100000	1.0	(100000)
1	36000	0.909	32728
2	36000	0.8264	29750
3	36000	0.7513	27047
4	36000	0.6830	24588
5	36000	0.6209	22352
Total cash flow	80000	Total NPV	36465

Total cash flow +80,000 Total NPV (net present value) +36,465.
From the example shown in Table 6.4, it is seen that all projects provide the same net cash inflow, but Project A, where the greatest influx of cash is at the beginning, has the highest present day value, and on this basis would be the one selected.

Internal rate of return (IRR)

This is a variation of the net present value method except that it is used when the cost of supplying the capital is unknown or uncertain. It is particularly useful for indicating the most profitable of several alternative projects. It requires the same data as the NPV method, except that the start point is an assumed 'break-even' total NPV, i.e. when NPV=0: the method is then worked backwards, using trial and error, to find the discount rate which, when applied to the annual cash flows, produces the break-even result. The discount rate thus arrived at is the internal rate of return, and on this basis the project showing the highest rate is the most profitable one. These computed rates of return are then compared with the enterprise's existing rate of return on its present investments, or against the cost of providing funds, to assess whether or not the new project is worth undertaking. Using the previous examples, it can be shown the internal rate of return for the three projects is:

Project A   27%
Project B   20%
Project C   22%

A summary of the NPV and IRR methods follows.

- Both methods indicate whether a project is acceptable or not compared with the minimum acceptable rate of return or the expected finance cost of funds applied.
- Both methods indicate a preferential ranking of alternative projects, on the basis of cash flow (NPV method) or profitability (IRR method) respectively. The higher the rate of interest applied, the less valuable are cash inflows received later, and the lower the impact of cash outflows incurred later.
- The NPV method assumes that the net cash inflow generated during the course of a project is reinvested at an interest rate no lower than that used as the discount factor. The IRR method assumes that the net cash inflow generated is reinvested at the IRR.
- The IRR method produces problems of computation where the cash flow pattern is irregular: that is, when cash outflows occur at future times in between the normal cash inflow occasions.
- Both methods involve an assessment of the cost to the enterprise of the capital it will use for investment. This in turn requires determination of the source of the funds to be used.

Another way of discounting is the annual equivalent cost (AEC). Whereas the NPV is the amount by which any future benefits at present value exceed the cost of the project, the AEC is the average amount by which the projects exceeds this in each year of the project's lifetime.

Sensitivity analysis

This is the process by which key design features are tested to determine what impact they may have on the project. The first step is to identify those components of the capital cost for which there is a margin of error, such as installation cost, or the variable cost of components sourced abroad. Sensitivity tests may also be applied to the benefits, such as errors in the estimate of post-implementation costs, particularly those caused by variations in fuel cost or the weather, or the lifetime of the measure.

Financing the investment

A range of investment routes is available, including investment from within and borrowing from banks. A further possibility is equipment supplier finance, in which a third party buys the equipment (e.g. a CHP plant) and takes on the responsibility of having it designed installed, operated and maintained. This is often considered an attractive option since it removes some of the risk, but inevitably this will be reflected in the cost of the energy supplied.

7   Legislation and grants

Introduction

Following the signing of the Climate Change Agreement in Paris in December 2015, it is anticipated that, worldwide, there will be a range of approaches to limiting carbon emissions. In December 2015,195 countries adopted the first ever universal, legally binding global climate deal at the Paris climate conference (COP21), which set out a global action plan to avoid dangerous climate change; it is due to enter into force in 2020. Governments agreed a long-term goal of keeping the increase in global average temperature to well below 2°C above pre-industrial levels and to aim to limit the increase to 1.5°C. Further, governments agreed to meet every five years to set more ambitious targets, report on how well they are doing to implement their targets, and devise a robust and transparent system of accountability. Richer countries would provide funding for poorer or developing countries to cope with the costs of complying. The agreement was signed on 22 April 2016 but will only come into force after 55 countries, accounting for at least 55 per cent of global emissions, have ratified it. Countries will set nationally determined contributions (NDC) which will determine that country's target. The agreements has been hailed as a milestone in international cooperation over climate change, but there has been criticism that the contributions are not binding, there will be no mechanism to force a country to set a target by a specific date, and no enforcement if a set target in an NDC is not met.

Although much of current UK government thinking seems to be to leave most of the activity to market forces, there will inevitably be legislation passed to promote the decarbonisation of energy. Historically, governments throughout the world have adopted three main strategies to reduce energy consumption and carbon emissions – encouragement in the form of grants, low or interest-free loans and tax incentives; compulsion in the form of regulations aimed at limiting heat loss and gain; and taxes on fuels. Since many of these measures vary from one country to another and are amended at regular intervals, it would be fruitless to go into great detail here, and only a brief outline is presented indicating typical schemes and the way they work. Surveys have shown that the main incentive to reduce energy consumption is cost. However, this alone appears to be insufficient to deliver the carbon reductions required, and stronger measures have been taken to encourage or force people to reduce their fossil fuel consumption. A number of schemes developed by the UK government are described briefly below; similar schemes are in place in a number of other countries.

A typical UK government scheme known as Affordable Warmth began in 2013 and replaced a similar, earlier scheme known as Warm Front. There are many conditions applying, for example these grants are available only for people in receipt of certain state benefits. For applicants who meet the criteria, assistance is offered towards new boilers, cavity wall insulation and loft insulation. It is due to finish in March 2017.

Feed-in tariffs (UK)

At present there are no direct grants from central UK government for the installation of solar photovoltaic (PV) panels, although some local authorities may have schemes in operation. Feed-in tariffs (FITs) are designed to promote the use of solar PV and other renewable electricity sources, by paying the owner not only for power exported to the grid, but for the power they generate and use themselves. The system is administered by the Office of Gas and Electricity Markets (OFGEM). Those eligible must have systems installed by an accredited installer. The tariff is paid for all the electricity generated, and varies depending on the source and the scale of generation; it also includes (at a much lower rate) existing generators transferred from the older renewable obligation (RO) scheme. The tariffs initially set were regarded as excessively generous, and have been reduced on two occasions, which will inevitably make home-mounted PV schemes less attractive. The scheme also covers wind, micro-CHP, hydro and anaerobic digestion. Smart meters will be needed to measure the amount generated, used and exported, and systems have to be registered with OFGEM. Several other countries have corresponding FIT schemes.

Renewable heat incentive (RHI)

Although 47 per cent of greenhouse gas emissions in the UK are attributable to heating, only 1 per cent of heat energy comes from renewable sources. The government's plans require that by 2020, 15 per cent of energy should come from renewables, and it has been estimated that they could supply 12 per cent of the heat demand. The RHI is designed to encourage the production and use of renewable heat sources by providing funds to owners for every kWh produced. The sources funded (at different rates) include biomass, ground source heat pumps, air source heat pumps and solar thermal. A premium will be paid for heat exported, but it must be demonstrated that the source is connected to a heat network – otherwise it would be possible to collect income by generating unwanted heat and wasting it – and certain quality standards must be met. The tariffs payable vary depending on the type of renewable and size.

Example

A 6 m^2 roof-mounted solar thermal panel in the north of England costs around £6,000 and produces 2,705 kWh/year. Under RHI the owners can claim 18 p/kWh = £2,705 × 0.18/year = £486.90 per year for 20 years, i.e. a total of £9,738.00, greater than the cost of the panel. This is in addition to the savings made by offsetting fossil fuel use.

These amount to (assuming gas used at 80 per cent efficiency and 4.5p/kWh):

£2,705 × 0.045/0.8 = £152/year
= £3,040 over 20 years, a total saving of £12,778.

Payback period (PBP) without RHI = 6,000/152 = 39 years.
PBP with RHI = 6,000/638 = 9.4 years.
Carbon dioxide (CO_2) savings are 498 kg per year.

With RHI and a ground source heat pump (GSHP), the example in Case study 2 (see page 139) would cut the PBP of the GSHP from 11 to 5.2 years.

Building regulations

Most developed countries either have or are producing building codes or regulations to limit heat transfer through individual building elements or to restrict the overall energy consumption of a building. Part L of the Building Regulations in England and Wales (Section 6 in Scotland) lays down the required standards for U-values, air-tightness, boiler efficiency, heating and hot water systems and controls, metering, and light fittings. Calculations are required in order to demonstrate compliance. Although many of the regulations only apply to new-build, increasingly stringent standards are being applied to refurbishments, extensions and conversions. Some of the practical implications are as follows. Replacement windows are required to conform to the new U-value standards (1.4 W/m^{2}K), implying the use of low emissivity glass instead of plain glass. The thermal performance of an extension will normally have to meet that of new-build. The required U-values at the time of writing are as shown in Table 7.1.

These values indicate the nominal maximum U-value allowed, but variations are permitted in certain circumstances – the reader is referred to the

Table 7.1 Maximum allowed U-values for new-build under the UK Building Regulations

Element	U-value $Wm^{-2}K^{-1}$
Walls	0.18
Roofs	0.13
Ground floors	0.13
Windows	1.4

relevant building regulations documents. Air leakage targets are also becoming stricter – designers claiming a maximum leakage rate of 10 $m^3h^{-1}m^{-2}$ at an applied pressure of 50 Pa will be required to demonstrate that the building meets this level by using the blower door test described in Chapter 5.

The regulations are reviewed at regular intervals and also now make reference to the reduction of cooling energy to encourage designers to introduce such passive systems as night cooling and increased thermal mass for damping of heat gains.

European directive on energy in buildings

This Europe-wide initiative came into force in January 2006, affects both domestic and non-domestic buildings, and was introduced to help meet Kyoto commitments and reduce energy consumption in buildings across the EU. Amendments were made in 2010. The main manifestation in buildings has been the display of energy performance certificates (EPCs), which have to be produced by a certified assessor. The Building Research Establishment (BRE) and others offer courses through which engineers and others can become certified. The calculation method used in the UK is the simplified building energy model (SBEM).

Main requirements

- Minimum energy performance for all new buildings (calculated by a prescribed method).
- Minimum energy performance for large existing buildings subject to major renovation.
- Energy certification for all buildings.
- Regular mandatory inspection of boilers and air-conditioning systems in buildings.

Public buildings are required to display the energy certificate, with the intention of encouraging greater energy efficiency.

Energy Performance Certificates (EPC) and Display Energy Certificates (DEC)

Documentation must be generated for new-build projects in order for the project to receive completion sign off. An SBEM must be provided to the building control officer In order to achieve building regulation compliance. From this model a Building Regulation UK Part L (BRUKL) report is generated, clearly defining and displaying a building's 'Building Emissions Rating' (BER) versus its 'Target Emissions Rating' (TER) for the individual building type. A building's BER is expressed in $kg/CO/m^2/year$ and is calculated from any energy used for electrical and lighting systems, ventilation, heating and cooling based on pre-defined occupancy levels. This document is not really meant as a prediction of the actual energy consumption for the building in question; it is merely a means to ensure compliance with the relevant section of the building regulations. Some have criticised this process on the grounds that it often renders the document effectively redundant as it could merely be a box-ticking exercise.

EPC

Energy Performance Certificates were introduced through the EU Energy Performance of Buildings Directive (EPBD) and provide only a simplified visual rating of how well a building should perform in relation to energy consumption. A figure for kilograms of carbon per m^2 of a building's footprint is used as an indication of its impact on the environment. The certificate uses a rating system A–G, an A rating being the lowest emissions and therefore the best. There is an additional feature that allows designers to give an indication of how certain improvements could produce better results. Examples of such improvements include renewable energy sources such as PV systems or insulation around domestic hot water vessels, and estimated payback periods are presented. The limitation of this approach is that there is a lack of detail and nuance in the approach, and that it does not take sufficient account of individual circumstances.

DEC

Display Energy Certificates apply to public authority buildings providing a public service such as an education facility with a floor area over $1000m^2$. A DEC provides a similar rating to an EPC (A to G banding) however the calculation process does not take place until the building has been occupied using a year of metered energy use.

The document is also useful in identifying poor energy management in plant such as air-conditioning units which over-cool certain spaces.

Energy Efficiency Directive

A further piece of European legislation is the Energy Efficiency Directive (2012). The requirements of this are:

- EU countries are required to make energy efficient renovations to at least 3 per cent of the buildings owned and occupied by central government.
- EU governments should only purchase buildings which are highly energy efficient.
- EU countries are required to draw up long-term national building renovation strategies to be included in National Energy Efficiency Action Plans, which member countries must draw up every three years. A number of support mechanisms have been put in place to enable countries to fulfil these requirements.

ESOS (Energy Savings Opportunity Scheme)

This is another part of the UK government's response to the EU energy directive, aimed at reducing carbon emissions by 20 per cent by 2020, and is one of a long line of mandatory energy efficiency initiatives developed by various government agencies. It includes energy used in buildings, industry and transport. The aim is to use auditing for firms to begin to manage their energy use. However, at present there is no requirement placed on them to act on the information generated and the scheme is limited to large firms, that is, organisations with 250 or more employees, an annual turnover of £39 million and a balance sheet of £33.5 million. Organisations will have to appoint a 'Lead Assessor' to manage the scheme and submit a report to the Environment Agency. They will have to show that at least 90 per cent of the energy use has been audited, and that ISO 500001 is being implemented. The lead assessor may be an in-house member of staff or externally appointed. ESOS assessments must be carried out every 4 years, and failure to comply will incur fines.

Climate change levy

The Climate Change Levy (CCL) came into force on April 1 2001, following the   recommendations made in Lord Marshall's report *Economic Instruments and the Best Use of Energy* (October 1998) and two years of consultation with industry. It is a tax on energy use in industry, commerce, agriculture and the public sector. All UK businesses and public sector organizations pay the levy via their energy bills, but fuel oils do not attract the levy as they are already subject to hydrocarbon oil duty. The effect has been to increase energy bills by 10 per cent or more. There is an additional £150 million of government assistance to business for energy efficiency measures. CHP may be exempt or attract less CCL compared with conventional generators, depending on a number of factors including the scale of generation, the quality index (QI), and power efficiency (PE) (see Chapter 4). Energy from renewable sources and cogeneration was originally exempt from the levy but this exemption has now been withdrawn, and energy-intensive users who sign a climate change agreement (CCA) with the Department for Energy and Climate Change are eligible for a reduction of up to 80 per cent. Small businesses paying VAT at a reduced rate are automatically exempt from the levy. One of the original intentions was to aid manufacturing industry by raising the price of energy relative to labour, and it was therefore expected to have a favourable impact on employment. The revenues raised are recycled to business through a 0.3 per cent reduction in employer's National Insurance contributions. Overall, the levy is intended to be revenue neutral: in other words, the amount the government gains in levy is paid in out in a reduction in the NI contributions. There have, however, been criticisms from manufacturing industry that some sectors lose out considerably, and that increasing energy costs have made British industry uncompetitive. Businesses that use large amounts of energy but have few employees (such as Scotch whisky distilling) are net losers, while employers that use relatively small amounts of energy but have a large number of employees, such as the Royal Mail letter services, are large gainers. Heavy industrial energy users can obtain partial rebates by entering into energy efficiency agreements with the government, and enhanced capital allowances are available to offset the effects to some extent. The CCL in the UK is administered by HM Revenue and Customs, to which application should be made for registration and exemption.

Emissions trading schemes

European Union's ETS (EUETS) covers over 10,000 industrial installations and power generation plants, which together are responsible for almost half the EU emissions of CO_2. It is a fundamental element of the European Union's policy to combat climate change and is at present the largest international system for trading greenhouse gas emission allowances. Its aim is to reduce industrial greenhouse gas emissions in a cost-effective manner and encourage energy efficiency. The system sets a cap covering the whole of the EU on the total amount of emissions of selected greenhouse gases, particularly CO_2, and the aim is to reduce the cap gradually over time so that by 2020 the emissions will be 21 per cent lower than in 2005; by 2030, it is proposed that they should be 43 per cent lower. Companies receive or can buy emission allowances which can be traded with other companies as needed, and as there is a limit on the total number of allowances available, it is ensured that they have a market value, and credits may be traded on an exchange. If a company's emissions exceed the allowances it has, fines are imposed. If it reduces its emissions, it can keep the spare allowances to cover future needs or sell them to other companies that need more allowances. The scheme is now in its third phase, which places a single EU-wide cap on emissions, as opposed to the earlier caps on individual nations; auctioning becomes the default method for allocation of allowances; and further allowances set aside to fund the deployment of carbon capture and storage technologies.

As an example of how emissions trading works, consider two companies, company A, which is able to cut carbon emissions at a cost to itself of £5/tonne CO_2, and company B, which is able to cut emissions at a cost of £9/tonne.

Company A sells 1,000 tonnes to company B at £7/tonne, i.e. £7,000, thus making £2,000 profit. Company B buys 1000 tonnes from company A at £7/tonne, saving itself £2000.

The rules apply mainly to those in the heavy industrial sector, such as iron and steel, cement, glass, ceramics and paper manufacture. Large emitters must monitor and report their emissions, and must return emission allowances equivalent to their emissions, on a yearly basis. It was hoped that the existence of a market would drive emissions downwards, and although some reductions in overall emissions have been made, it is felt that more could have been achieved if targets had been made more demanding. One criticism of the scheme is that much of the reduction in emissions that has taken place has been attributed to the economic downturn rather than improvements in efficiency. It has been suggested that one result of the 2015 Paris Climate Change Agreement could be a global emissions trading scheme.

Carbon reduction commitment (CRC) energy efficiency scheme

The CRC scheme is aimed at cutting those CO_2 emissions in large public and private sector organizations not already involved in CCAs under the CCL and the EUETS. The organizations involved are collectively responsible for about 10 per cent of UK emissions. For organizations that consume more than 6,000 MWh per year, metered through a half-hourly electricity meter, participation is mandatory, and they must register with the Environment Agency. Annual league tables of performance are produced, and organizations are encouraged to develop better energy management strategies. The scheme began in 2012, and participants will buy allowances from the government each year to cover their emissions in the previous year. Thus if they reduce their energy use they can lower their costs. Allowances were initially to be priced at £12 per tonne of CO_2 for the first year and then allowed to vary according to market demand. Prices are expected to be in the range £16.10–£16.90 per tonne for 2015/6.

Education, training and recognition

With the greater awareness of energy issues and increased legislation on energy use, the demand for energy managers is only likely to increase in coming years, and along with this growth there has been an increase in the provision of education and training in the energy sector. Many university courses on traditional subjects such as mechanical engineering, electrical engineering, etc., now place greater emphasis on energy matters and the importance of energy conservation, while a number of specialist energy degrees have emerged, both at undergraduate and postgraduate level. Architecture courses are also doing more to promote energy-conscious and sustainable building design, and often include the study of advanced building modelling techniques. In the UK, accreditation of a number of these courses by bodies such as the Energy Institute (EI), the Institution of Engineering and Technology (IET) and the Institution of Mechanical engineers (IMechE) enables successful energy professionals to become chartered engineers, and the EI also offers Chartered Energy Manager status. Details about these courses can be found on the EC-UK website and also on those of the accrediting bodies.

Not all those involved in energy management need necessarily be professional engineers, and other training and education options exist; organisations such as EI and the Energy Managers Association provide energy management training and combined heat and power (CPD) at a range of levels, from basic courses for those just entering the industry, to advanced courses for experienced professionals. Details of these can be found on their respective websites. The carbon trust produces a large number of publications related to all aspects of energy management in buildings, and workshops, often free, held throughout the UK. Contact details for these organisations can be found in the References section of this book.

8   Controls and building management systems (BMS)

Heating, ventilation, lighting and air-conditioning systems all require controls so that the spaces they service enjoy the environmental conditions demanded. Controls are often, and perhaps should be, almost unnoticed by the building user, and automatic control can be a contentious issue. In a modern building it is expected that appropriate conditions will be provided without the need for intervention but the solution is not to simply force automatic controls on every situation; occupants like to feel they have some say in the conditions in their workplace, and productivity and satisfaction drop considerably when all control is taken out of occupants' hands. It has been found that building users who are able to have some control over their environment tend to adapt better to more extreme conditions; the design conditions for naturally ventilated or mixed-mode buildings with user control are thus often less rigid than for those that are fully air-conditioned. This can be the cause of conflict, and the optimum solution is often some form of automatic control, but with manual override facilities to suit the user's needs. The problem for shared working environments is to maintain comfort for the majority of people while still catering for individual needs.

Comfort parameters for winter and summer should be set appropriately, and the acceptable temperature ranges are often beyond those that have traditionally been used. For example, summertime temperatures above 25°C for up to 5 per cent of the working year and above 28°C for 1 per cent of the working year have been found to be acceptable, and would save over 10 per cent of cooling energy compared with an absolute limit of 25°C for the whole year.

Conventionally, each service had its own control system, but in the last 30 years, BMS, which can in theory control all the systems under one umbrella, have become standard in most non-domestic buildings. BMS are also known as building automation systems (BAS), energy management systems (EMS), energy management and control systems (EMCS), central control and monitoring systems (CCMS), and facilities management systems (FMS). For the sake of simplicity, the term BMS is used here to cover all these variations. Before BMS were developed, individual modules were used to control specific functions or items of building services plant;

they are still used in smaller buildings and of course in many older buildings. These typically control:

- temperature
- humidity
- time-of-day switching
- power management
- optimum start/stop
- light switching.

They have certain disadvantages in comparison with BMS. They are not necessarily much cheaper, the data is not available for output elsewhere, and they tend to operate on fixed algorithms with no centralized control and no monitoring and logging facilities. While a dedicated controller for an air-conditioning system may control the temperature and humidity well, it has certain limitations from the point of view of energy efficiency:

- it may not be possible to have a complete overview of all the conditions and settings at any one time
- there are generally no facilities for recording settings, environmental conditions or operation, to build up a historical record of performance
- changing switch-on and switch-off times may be a cumbersome process requiring manual intervention at the plant itself.

These limitations make it difficult to assess the energy efficiency of the system or to tune it for maximum efficiency, but they can be addressed by the use of a BMS. The cost of the energy consumed in non-domestic buildings in the UK is about £8,000 million per annum, rising steeply year by year. BMS have the capability to operate buildings more efficiently to reduce these costs, but in many instances the operators are non-technical personnel with a limited understanding of their potential. A correctly set-up and operated BMS can help with automating the operation of a building and improve energy efficiency, and one that is incorrectly used not only costs a great deal to purchase, but could actually waste energy. Raising staff awareness on the operation of the BMS and energy management generally is important, as has been mentioned elsewhere.

The operations that a BMS can perform range from the simplest task, such as turning off heating when not required, to the performance of complex procedures, such as tracking temperatures, adjusting outputs, and logging control sequences. The latter can be particularly useful in fault-finding and can save a great deal of time. In the hands of a well-trained operator a BMS enables energy usage to be firmly controlled. As well as controlling those functions mentioned above, it might also interface with other control and monitoring services such as safety, access, lifts, security systems and fire security systems, or even be fully integrated with them. A BMS may control only a few functions in a small building, or many functions in a group of buildings many miles apart. It is important that it is no more complex than required for successful running of the building, because excessively complex systems are a waste of money and tend to be underused. Only a few years ago BMS were only considered as an afterthought for many buildings, but if they are designed-in from the beginning of the services-design process they should enable greater efficiencies in both energy and operational costs to be achieved. It is possible to retrofit BMS but cabling, etc. can prove difficult, and may lead to additional costs, and it is often the case that it is not worthwhile incorporating a BMS into an older building unless a general refurbishment is taking place. Careful estimates of the costs and benefits are therefore essential in such situations. In the past, BMS were only cost-effective in very large buildings, but in recent years the cost in real terms has fallen so much that they are competitive with stand-alone controls in most buildings. Current systems are modular in form, which means that if the building is extended, units can be linked together to control the enlarged building. In a hard-wired system, much of the cost is associated with cabling and installing the sensors; wireless sensors and actuators are now available, but at a higher cost. Sensors are an essential feature of a BMS, and wiring-in the sensors may constitute a substantial proportion of the installation costs of a system. In addition, retrofitting sensors to an existing building may present many difficulties due to its internal layout and construction. Wireless sensors may help to overcome some of the difficulties which emerge, and the use of scavenging sensors, which use heat in the room to operate, can avoid the problems of battery replacement. Some systems also use wireless actuators. There are, however, difficulties in implementing wireless systems, such as loss of signal in buildings with high concrete mass, interference from other signals from phones, Wi-Fi etc., and so it is not always the ideal solution. Future-proofing of systems to allow for upgrades, expansion and change of use is also important in the initial specification. Correct maintenance of a BMS system is important, particularly regularly updating set-point settings, hours of operation, etc. Sensors can drift out of calibration, particularly humidity sensors, and periodic checks should be carried out. Changed occupancy levels, moving partitions around within an open plan space, or moving items of equipment around can sometimes adversely affect the operation of the BMS and heating, ventilation and air-conditioning (HVAC) system – for example, inadvertently locating a photocopier, which emits heat, in front of a sensor controlling the temperature of a space. Operational expenditure needs to be considered too, as it is not always a trivial amount. The cost and time involved in analysing data may be considerable on a large site or on an operation controlling a number of buildings.

Comfort and environmental control

Heating and lighting systems have long been subject to automatic control, since it is neither convenient nor cost-effective to continually adjust controls on boilers and air-conditioning systems in large buildings. A major stage in automatic control came with the development of central heating provided from a single boiler controlled by a suitably located thermostat. Temperature control of individual rooms became available with thermostatic radiator valves (TRV), which can be manually adjusted to suit individual requirements. The use of gas and liquid fuels, along with electric timers, made automatic control of the boiler easier and provided some flexibility in timing, but it was the advent of physically small and cheap computing systems that heralded the development of BMS with fully flexible timetables and multiple zones, making possible the associated managerial, operational and energy efficiencies.

Reducing set-points for heating and raising them for cooling systems is an easy way to reduce consumption, but thermal comfort should not be adversely affected. A compromise generally has to be reached between completely individual controls and central control. The former can sometimes lead to widely fluctuating temperatures as a number of individuals try to control systems to suit themselves. The ease with which thermal comfort can be achieved depends not only on the systems installed but on the characteristics of the building itself. Buildings can only provide thermal comfort and low energy consumption simultaneously if they can respond readily to the changing requirements of the occupants

Ensuring that thermostats or temperature sensors are fitted in the right place is critical. Avoid locations close to draughts or heat sources as they will not give a representative room temperature. What is often forgotten is that if a building's use changes, or there are internal rearrangements or partitions or furniture, then the optimum thermostat location may also change. Smart valves enable radiators to be switched off at pre-determined times, and in addition TRVs are available with PAIR (passive active infra-red) sensors which enable the heating automatically to be reduced when areas become unoccupied.

Two-position control

In order to understand the benefits of a BMS, it is useful to consider a simple stand-alone control system, an example of which is shown in Figure 8.1. Here a room is provided with ventilation at a constant rate, and a thermostat temperature sensor switches a heater battery on and off in order to maintain a constant temperature. When the thermostat detects a deviation below the set-point temperature, a signal is sent to the actuator in the controller to open a switch that turns the heater battery fully ON (using a two-position ON/OFF control). The information from the signal sent by the thermostat is available only to the actuator, and the switch is controlled only by the information from it. This limited use of information is typical of conventional controls. The only other information controlling the system might be from an overriding timer which limits the ON and OFF periods to

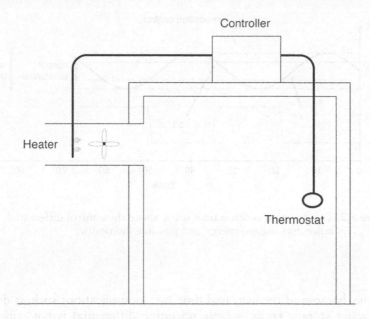

Figure 8.1 Conventional control of a single-zone constant-volume all-fresh air
ventilation system using two-position control

certain hours in the day, and switches the fan off and on at those times. A
cut-out link prevents the heater from operating when the fan is off. The
simple control system in Figure 8.1 is not especially effective at giving close
control of temperature in the space, and is not optimized for energy efficiency. Nor is there the facility for recording the conditions, running hours
and energy consumption of the system. The two-position control (ON/OFF)
is particularly poor, and is dependent on the electromechanical properties of
the components. Solid-state systems are capable of providing much more
accurate control of conditions.

The thermostat used to control the heater battery in Figure 8.1 is likely to
be of the bimetallic strip type, and switches the heater battery fully ON or
OFF. A control differential defines the limits between the on and off points
on the thermostat; if it is too small the switching frequency will be too high,
leading to excessive wear and accelerated breakdown; if it is too large the
swing in temperature will be too great. An operating differential defines the
difference between the highest and lowest temperatures in the room as the
heater cycles between on and off, and is larger than the control differential
because of the time lags in the system. After the heater is turned off, its thermal mass causes it to remain hot for some time, allowing the temperature to
overshoot on the high side of the set-point (see Figure 8.2). The greater the
physical distance between the heater and the room, the higher the overshoot

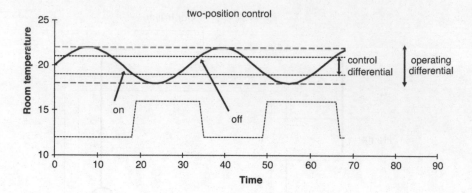

Figure 8.2 Two-position control: time spent above the control differential
represents wasted energy and possible discomfort

will be, because of the increased time lag. In applications such as domestic
hot water storage tanks, a large operating differential is not critical, but
when heating a room any overshoot represents wasted energy and possible
discomfort for the room occupants.

The operating differential can be reduced by employing timed two-position control, where a small heater is built into the thermostat. When the
thermostat calls for heat, this heater is energized and the heat generated
within the thermostat causes the thermostat to close the heating valve
earlier and limit overshooting, resulting in a small improvement in performance.

Proportional and other controls

In contrast to two-position control, which is 100 per cent ON or 100 per
cent OFF, many controls, particularly within BMS, are proportional. That
is, a control unit sends a signal to the heater whose magnitude is proportional to the deviation from the set-point. As the temperature approaches
the set-point, the output from the heater falls, so overshoots are reduced,
and both temperature control and energy efficiency are improved. In fact
the temperature never reaches the set-point as the output from the heater is
dependent on the difference between the sensed temperature and the set-
point, and at very low differences the heater output is negligible, so there is
always a small temperature offset. Other forms of control such as floating,
proportional, integral, or derived-control perform better than two-position
control, and are found in BMS applications. PID (Proportional-Integral-
Derivative) controllers allow the faults of each type of control to be
minimised and are common in HVAC systems.

Closed and open loops

The system shown in Figure 8.1 is an example of a closed-loop system in which the result of the control action (that is, the room temperature) is fed back to the controller. In an open-loop system there is no feedback and the result of the control action does not affect the input to the controller. An example of this is heating control based on a thermostat mounted on an outside wall of a building. Its operation is based on the assumption that the heating demand is inversely proportional to the outdoor temperature, which in a general sense is correct, but it does not take into account a number of other factors that influence the heating demand, such as open doors and windows and variations in the number of occupants. The outdoor temperature is unaffected by either the state of the heater or the indoor temperature. Clearly, such systems used alone have serious limitations and may result in poor control of internal conditions, but they may still be found in older buildings. Often, though, outside temperature sensors are used in addition to indoor sensors in a more complex control-loop – for frost protection, for example, or in a compensator. An example of where an open -loop system can be useful is a lighting control system where the lights can be dimmed according to the outside light level. The level of daylight outside the building will not be affected by the lighting level inside the building. Open-loop control works in this instance because there is a fixed relationship between the indoor and outdoor lighting levels, based on the orientation, window size and window position.

System compatibility

In the early days of BMS, components from different manufacturers were not compatible, because of the use of proprietary communications protocols (the 'language' used for communication within the system). The implication of this was that the whole system – sensors, actuators, central control and software – had to be purchased from the same manufacturer. More recently, movement towards a truly common communications protocol has been made, and a number of 'open protocols' have been available for some time, including BACnet, Batibus, Lonworks and EIBus. Thus a BACNet-compatible valve actuator from one manufacturer will operate correctly with a BACNet-compatible sensor from another. KNX is a standard open interface which allows the integration of a number of applications, for which around 200 manufacturers produce compatible devices. Lighting control is often run from a completely separate system from HVAC, and a number of sophisticated lighting controllers use the DALI (Digital Addressable Lighting Interface) protocol, which is KNX compatible. Fire and safety systems may be integrated with the BMS, but it is often felt that a separate fire safety system gives greater protection than one that is fully integrated. If the systems do not use the same protocols they can be integrated through 'gateways' installed for each of the separate services, which can be programmed to provide a communications network between the separate systems.

Functions of a BMS

The functions provided by a BMS include control of plant, such as:

* automatic switch on/off of heating, ventilation, air-conditioning and lighting
* optimization of plant operation and services to minimize energy consumption and improve maintenance
* maximum use of outside air for cooling air-conditioned buildings
* the provision of multiple timetabling and scheduling opportunities, for example weekdays, weekends and holidays
* implementation of, for example, Uninterruptible Power Supplies (UPS) in the event of power failure and so on, in critical environments such as a hospital operating theatre, etc. Spare controllers/sensors etc. or redundant systems may be installed as back-up ready to take over.

Functions concerned with monitoring of plant status and system variables include:

* sensing values of important parameters such as temperature, flow rate, humidity, valve position, and plant on/off status
* generating alarms when pre-set values are exceeded
* helping with maintenance by assessing the state of plant (such as the cleanliness of filters)
* taking more rapid remedial action in case of faults, thereby minimizing damage or disruption
* additional functionality may include default to latest working configuration in the event of power loss or software crash
* controllers logging data locally if main communication with the server is lost.

A BMS can be used as an aid to effective maintenance. Routine maintenance increases the potential life of a control system and reduces the frequency of breakdowns and emergency repairs. The output from the BMS can be linked to a computer-based maintenance system which enables preventive maintenance to be carried out, making it easier for replacement to be made based on the number of hours plant has run, rather than using a fixed period of elapsed time. The switching functions of items of plant such as compressors, fans, pumps and boilers can be logged, and the software will total the hours run for each item of plant in order to generate appropriate work orders. This creates an effective preventive maintenance regime under which replacement of parts can be carried out at convenient times with a consequent reduction in breakdowns and unplanned timeouts. It may in fact lead to the amount of maintenance work being reduced, as the servicing is scheduled according to the hours run rather than time elapsed. When an item of plant such as a pump goes beyond its recommended service interval, an alarm report can be generated to remind the staff that it is overdue for servicing. Maintaining items of plant in top condition will enable them to run at their maximum efficiency for longer and save energy in the long run.

A common BMS feature is the use of pressure sensors to monitor the pressure drop across an air filter in a HVAC system. As the filter becomes dirty the pores clog up and the pressure drop across it increases. When this pressure drop exceeds the set-point the BMS can be programmed to issue an alarm indicating that the filters need changing. Another form of break-down is when an actuator handle shears off the shaft on a three-way valve controlling a heating system: the valve actuator position sensor indicates that it has closed correctly, but the hot water continues to flow through the system and the room temperature rises beyond its set-point. Monitoring the temperatures and logging the sequence of control actions enables the fault to be traced more quickly than would be possible without the BMS. BMS also score over stand-alone controls in the provision of energy management information such as the availability of instantaneous energy consumption and temperature values. The extraction of stored historical data and comparison with current values enables trends in consumption to be identified, and data for M&T (monitoring and targeting), degree days, CUSUM and other forms of analysis to be generated. Over the longer term, this allows the effectiveness of energy-saving measures to be assessed by 'before' and 'after' interventions.

In Chapter 3 a method of assessing the heating system using data from a BMS was described. One major task in the UK is the twice-yearly changing of the clocks; for stand-alone plant every time clock must be changed manually, and experience has shown that often clocks on HVAC systems are not changed as they should be, resulting in inappropriate timing of heating and cooling periods. With a BMS, the clocks for an entire installation can be adjusted for summertime change by using a few strokes of the keyboard instead of visiting each plant room separately. Often BMS are set-up to produce a large amount of data, much of which goes unanalysed, and surveys suggest that only 30 per cent of the capacity of a typical BMS is currently put into use, largely because of insufficient training, while another survey of 50 managers showed that 82 per cent had BMS but only 2 per cent were able to use them for targeting. Better training in the use of BMS is therefore essential.

BMS configurations

A BMS requires sensors to measure variables such as temperature and pressure, actuators which will switch the plant on and off and vary the position

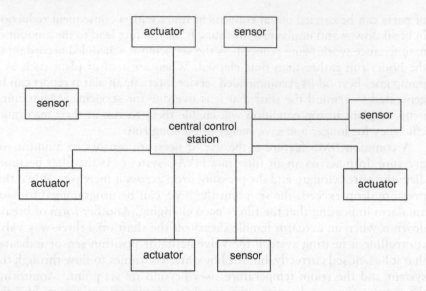

Figure 8.3 Star or radial layout of a BMS

of valves and dampers, and an intelligent controller. A number of configurations are possible.

A radial or star layout, typical of older systems, is shown in Figure 8.3. Each sensor and actuator is hard-wired directly to a central control station. This layout is simple and effective, but in larger buildings may require long cable runs, which are expensive and which may result in significant voltage drops, leading to operational difficulties. All the processing will be carried out at the central station. Newer systems use outstations which gather together local data and control points and permit shorter cable runs (Figure 8.4). Since each outstation possesses a certain level of processing capability, a central station is not necessarily required, as the outstations can be connected together. Access for downloading data or changing set-points can be made by plugging in a laptop, and can be controlled by the use of passwords. This has major advantages for large organizations such as local authorities and those with sites spread widely over a large area. A central 'building management facility' can control and collect data from all the buildings on the system into one location. It is common now for BMS to be accessible through internet browsers, providing even greater flexibility, and buildings on sites throughout the country can be monitored and controlled from a central facility. Interaction with other software is increasingly desired, through internet connection with Ethernet, 3G services and other IT functions of a business, in an integrated system. BMS operation is more and more an IT-based operation, and may include the use of smartphone apps.

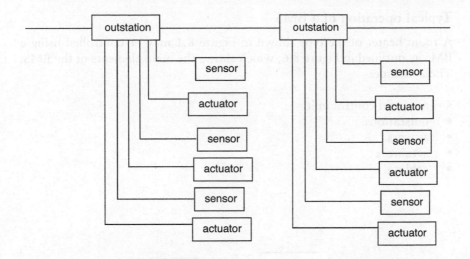

Figure 8.4 Bus connection of multiple intelligent outstations

Back-up and storage of data is important in the correct operation of BMS; back-up can be local, server-based or cloud-based.

A further BMS configuration is ring topology (see Figure 8.5), in which the network cable is connected to each station in turn and information travels round the ring in one or both directions.

This is often used in integrated systems, which may include intruder and access systems, fire alarm and file management systems along with the control of the HVAC equipment.

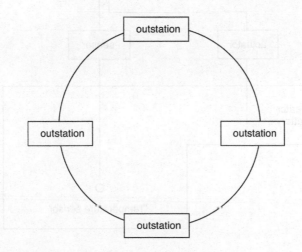

Figure 8.5 Outstations connected in ring topology

Typical operation of a BMS

A room heater of the type shown in Figure 8.1 may be controlled using a BMS as outlined in Figure 8.6, which shows the main elements of the BMS. These include:

- a central control unit
- outstations
- sensors
- actuators
- cabling.

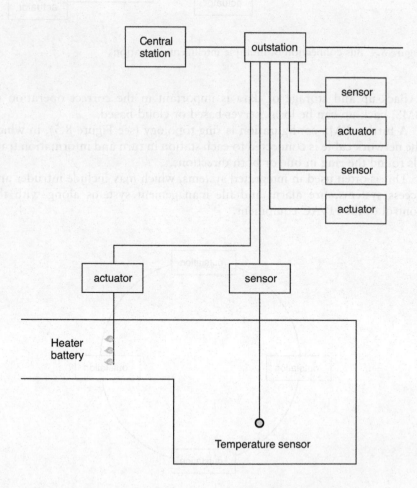

Figure 8.6 Temperature control as part of a larger BMS

A signal is sent from a temperature sensor in the room (such as a thermo-couple or thermistor) via the cabling to the outstation. At the outstation, the signal is checked against a previously entered set-point temperature. If the temperature is below the set-point a signal will be sent to the actuator to switch the heater battery to an appropriate level of output. Periodically, signals indicating the room temperature and the heater battery status will be requested by the outstation and central station, and will be transmitted and stored by the central station as required. In distributed-intelligence systems the central station will only be involved in setting the set-point and in recording the system settings; the major part of the processing is carried out by the outstation.

Compensated on/off control of a heating system

In compensated control the flow temperature of the water flowing through a central heating system is adjusted according to the outside air tempera-ture. If the outside temperature rises, the water flow temperature is reduced, thus allowing lower temperatures to be used in milder conditions (see Figures 8.7 and 8.8).

The temperature of the water flowing from the boiler (as sensed by the flow temperature sensor) is varied as a function of outdoor air temperature by the controller, which varies the position of the three-way valve and also

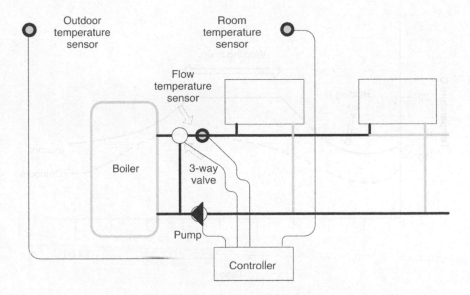

Figure 8.7 Compensated control of a heating system

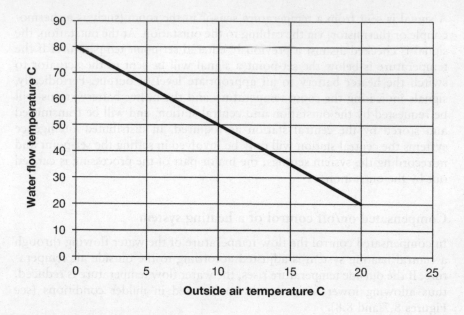

Figure 8.8 Compensator schedule

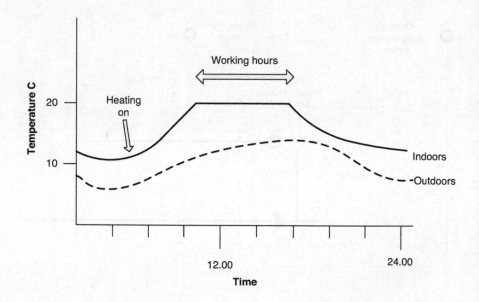

Figure 8.9 Optimum start

cycles the burner on and off. The room-temperature sensor may also be used to provide room influence control by resetting the compensation ratio. An increase in room temperature would lower the compensation ratio, and a decrease in room temperature would raise the compensation ratio.

Optimum start/stop

In most heating systems there is a certain amount of thermal inertia as a result of a physical distance between the sensors, heating element and control device (i.e. damper or valve). This is generally referred to as distance-velocity lag and it means that there is a delay in the system response to a change in conditions, and the effects of switching the heating system on or off are not felt immediately. In extreme cases it may take two hours or more for an office to become comfortably warm on a Monday after the building has been closed for the weekend. Similarly, at the end of the day the building will remain warm for a similar length of time.

In a heavyweight building, such as an old stone church, the time lag may reach several hours but in a small lightweight building it may only be a few minutes: it will heat up quickly and cool down quickly. Most buildings are of intermediate thermal weight and in order to achieve the required temperature at the appropriate time, the heating must be switched on some hours beforehand. How much earlier is important and depends on a number of factors, in particular:

- the temperature inside
- the temperature outside
- the thermal mass of the building and the type of heating system.

In older buildings without BMS, simple timers were set for the severest conditions and rarely altered subsequently. In milder conditions, which in fact cover most of the heating season, the building may reach the target temperature some hours before opening time, thus wasting energy (see Figure 8.9). There is a maximum rate at which the internal temperature can rise, depending on the thermal properties of the building and the heating system, and the outside temperature. Optimum start controllers incorporate both an ambient air temperature sensor fixed to the outside wall, and an inside temperature sensor. They include algorithms to calculate the appropriate start-up time from a combination of time and temperature inputs, and usually include a frost protection facility. The objective of optimum start is to ensure that the building reaches the required temperature exactly when required. Any earlier and energy is wasted, any later and the occupants will be uncomfortable for some time. Optimum stop control ensures that the building does not remain at a high temperature any longer than needed.

Optimum start/stop heating and air-conditioning controllers are available. UK Building Regulations specify that an optimizer or optimum start programme is required for buildings with a space heating load of more than 100 kW.

Electrical load management

Demand-side management is so called because customers create a demand for energy which is then supplied mainly by utility companies. In good energy management the correct specification of energy-efficient appliances and rational use of them enables demand to be minimised.

Supply-side management refers to the various operations carried out by utility suppliers, and is normally designed to enable them to supply power or fuel at the cheapest generation prices, and to set tariffs at a level that recoups the generation prices and provides an appropriate profit. As demand for electricity increases, less-efficient generating stations are brought on stream so the average cost of generation increases, and the price-per-unit paid by the customer increases accordingly. Per kilowatt hour, electricity is usually the most expensive fuel, and accounts for a substantial proportion of the fuel bill in office buildings, which have high air-conditioning and lighting loads. Peak-lopping and load-shifting operations may be programmed through the BMS to reduce high demand penalty charges and avoid peak charging periods.

Large users of electricity pay a maximum demand charge based on the maximum power used at any one time over a specified period, in addition to the actual electricity units used. For a large organization this can amount to several thousand pounds per month. Maximum demand penalties can be avoided by staggering machine start-ups, and shedding loads according to pre-set priorities: for example, switching off non-essential items of equipment at times of high demand, and so lopping the peak demand and reducing the over-charge. This does not actually save energy but reduces costs to the consumer. In items such as water heaters and refrigeration plant there is considerable thermal inertia, and switching off for short periods results in a non-significant loss of service and may avoid high-cost penalties. Demand can be monitored and plant switched off on a priority and size basis. Money could be saved by optimizing the agreed maximum demand amount. As part of this approach, it may prove cost-effective to incorporate energy storage, such as a large hot water tank, which will enable money to be saved by storing energy during low-cost periods and saving it for use at times of high cost. With renewable sources some form of storage is often essential in any case to make the system run properly and economically, and the cost of the storage must be taken into consideration. This may also be combined with a voltage optimisation system.

The example in Figure 8.10 shows how a staggered start-up of two electric motors enables the peak power drawn to be reduced. The start-up power is high, but demand falls back to a fraction of the peak value after a few seconds. If both motors are started simultaneously peak demand is 4 kW, whereas if the loads are staggered the peak demand falls to 2.5 kW.

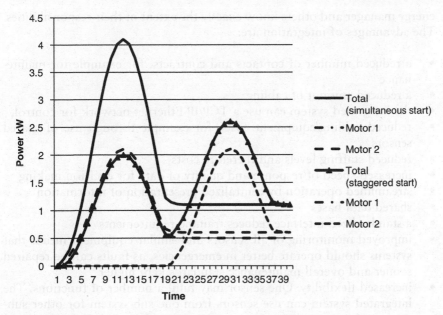

Figure 8.10 Effect on maximum demand of staggering switch-on

Savings from BMS

A number of sites have been surveyed to estimate where BMSs were achieving savings, the results being shown in Table 8.1. An overall payback period of 5.8 years was achieved. In one building 30 per cent of sensors gave dubious readings because commissioning was rushed.

A building, including BMS, fire, security, and access control, may be integrated in a number of ways, either physically or organizationally. While physical integration has been technically possible for some time, it can create organizational difficulties, such as a clash of functions. For instance, if security is integrated with environmental control, it may not be clear where responsibilities lie, and so is often not implemented. Detailed contractual arrangements need to be put in place from the outset so that the

Table 8.1 Savings from BMS

Measure	% saving
Increased heating system efficiency	6.6
Optimum start	5.1
Optimum stop	3.1
Correct holiday settings	4.1
Reduced internal temperature	4.6

energy manager and others know exactly the extent of their responsibilities. The advantages of integration are:

- a reduced number of contacts and contracts, for example for maintenance
- a reduced amount of cabling
- the integrated system can use a TCP/IP Ethernet network for control,
- reduced cabling/equipment costs, for example through use of shared sensors
- reduced staffing levels and therefore costs
- increased speed of response and quality of data for decision making
- streamlined operation by centralized presentation of information
- shared data bases
- a standardized interface reduces training requirements
- improved monitoring of all sensors and similar equipment means that systems should operate better in emergencies, as faults can be repaired sooner and overall maintenance standards kept higher
- increased flexibility. One sensor may fulfil a number of functions. The integrated system can use sensors from one sub-system for other sub-systems: for example, occupancy sensors used for lighting control can detect the presence of personnel in the building and alert the security sub-system. It may also be desirable to link this in with the fire alarm system, so that if the access system indicates people are working in a particular area, then in the event of a fire, those in charge know to look out for people there
- improved energy efficiency. occupancy sensors or access sensors can also be linked to the hvac system, so that rooms are only heated or air-conditioned when occupied.

Disadvantages or obstacles to integration are that:

- if one system fails, all systems may fail
- traffic loading on the network could cause problems. This is potentially dangerous if the fire alarm system is trying to communicate over an already busy network
- there could be disputes over job functions and seniority
- in spec-built offices the precise functions required may not be known
- there are always 'teething problems' when implementing new technology
- the interface might become too complicated for efficient operation
- downtime for maintenance might affect all systems simultaneously
- different protocols in the sub-systems might lead to excessive use of gateways and increase hardware costs.

Case study 1
An office building with medium-level glazing

This case study concerns an office building located in an out-of-town business park in central Scotland. It consists of mainly open-plan office space with a small number of meeting rooms. The rooms are spread over two storeys and the net usable area is 3,398 m². The heating is controlled by a thermostat in each open-plan space. An air-conditioning system cools the server room and the meeting rooms and conditions approximately 10 per cent of the floor area of the building. Operation of heating and cooling plant is based on the hours of 07.00 to 22.00 each day, the boiler efficiency is estimated at 70 per cent, a BMS is fitted but only the basic features are utilized.

The energy consumption figures are shown in Table CS1.1, and may be compared with the benchmark figures from ECG 19 in Table CS1.2. A limited energy audit was carried out on the building.

The figures show that although the performance is better than typical, it falls short of best practice. Thus, there is room for some improvements to be made. Closer inspection of the figures shows that there is considerable potential for improvement in heating, and some scope for improvement in lighting. The U-values of the elements, from the original building data, are as shown in Table CS1.3 and the glazing ratio (GR) is 50 per cent. The lighting level throughout the offices is 300 lux, which is an appropriate level for open-plan offices but is not particularly efficient at 20 W/m². This gives

Table CS1.1 A summary of the energy consumption, percent of total, and costs, for case study 1

Application	Consumption kWh	% of total consumption	Cost £	% of total cost
Heating	671,664	79.8	30,224	61.8
Cooling	6,770	0.8	744	1.5
Lighting	95,144	11.3	10,466	21.4
Other	67,960	8.1	7,476	15.3
Total	841,518		48,910	

Table CS1.2 Good practice and typical annual energy consumption for the building based on benchmarks in ECG 019 (adjusted to allow for A/C for 10% of the building)

Application	Consumption kWh good practice	Consumption kWh typical
Heating + hot water	329,606	604,844
Cooling	4,757	10,533
Lighting	91,746	183,492
Other electricity	101,940	132,522
Total	528,049	931,391

a total installed wattage of 69,960 W, which, allowing for a diversity factor of 0.5, gives an annual energy consumption for the lighting of 95,144 kWh. Other electrical use is estimated at 20 kWh/m²/year, giving an annual total of 67,960 kWh (11.4 per cent of total energy use). The building is located in Scotland where heating degree days = 2,500 and cooling degree days = 108. Applying correction factors of 0.75 for five-day week use and 0.6 for intermittent plant operation, the annual heating energy consumption = $17,413 \times 0.75 \times 0.6 \times 0.024 \times 2,500/0.7 = 671,644$ kWh/year (80 per cent of total energy use).

Allowing for a coefficient of performance (COP) of 3.0 for air-conditioning, the cooling energy is $17,413 \times 0.75 \times 0.6 \times 0.024 \times 108/3 = 6,770$ kWh/year (0.8 per cent of total energy use). A summary of the energy consumption, percentage of total, and costs, is given in Table CS1.1.

The lighting constitutes over 20 per cent of the total energy costs; improvements could involve better control and/or the use of more efficient tubes and luminaires. Short-term low-cost measures to improve the lighting could include stickers next to light switches. Longer-term measures requiring significant investment could include the use of LEDs, changing the

Table CS1.3 Case study 1 heating energy consumption figures

Element	U-value (W/sq.mK)	Area (sq.m)	UA (W/K)	% Fabric heat load	% Total heat load
Walls	0.84	480	403.2	5.5	2.3
Glazing	5.6	480	2,688.0	36.7	15.4
Roof	0.49	1,699	832.5	11.4	4.8
Floor	2.0	1,699	3,398.0	46.4	19.5
		ΣUA	7,321.7		
Ventilation conductance		(0.33nV)	10,092.0		58.0
		TLC	17,413.7		

lighting control strategy to include occupancy sensing or daylight-level sensing; it would be appropriate to instigate these during a comprehensive lighting replacement programme whenever that was due. The long-term strategy would include an investigation into the costs of such a programme. The heat loss from the curtain walling represents only 5 per cent of the fabric heat load and 2.3 per cent of the total heating demand, therefore any improvements in insulation to the walls are unlikely to prove cost-effective. A reduction of the U-value by 50 per cent, resulting in a lowering of the energy consumption by only 2.5 per cent, would save only £130 per year, at a cost of over £5000 – a payback period of over 40 years. This level of investment and payback period could not be justified. A change from single to double glazing saves £750 per year but with a payback period of over 20 years. This could be justified if there were other reasons for changing the glazing, such as rotting or corroded frames, or for purposes of soundproofing. At this particular out-of-town location, sound pollution is not a problem and the latter does not apply. The double glazing would have the further effect of reducing draughts and infiltration, and although taking this into account would reduce the payback period by another five to six years, it remains an unattractive proposition.

Specific recommendations

- If double glazing is not to be implemented, then some form of draught stripping would reduce infiltration, with a payback of three to four years.
- Addition of thermostatic radiator valves (TRV) would improve control and heating efficiency at a cost of £10–15 per radiator.
- A rolling programme of installing sub-meters could be paid for from the energy savings.
- Investigate the possibility of providing a separating chiller for the server room so that cooling operation can be optimized.
- Organize an information campaign to urge staff to turn off equipment such as computers, photocopies and printers at night.

The use of the BMS should be extended to include:

- optimum sequencing of boilers
- night set-back for heating
- optimum start/stop for heating
- limit basic hours of operation to 09.00–18.00 Monday to Friday, and instigate a booking system for provision of services outside these hours
- set the BMS to log performance and use the results for monitoring and targeting.

This would require some investment to train the building manager in the use of the BMS, but is likely to give very quick returns.

The long-term strategy should include:

- establishing a database of energy consumption figures and related records
- identifying the company's strategic plans and investment criteria
- establishing an energy efficiency programme
- establishing a replacement programme for lighting.

The specific recommendations for improvement, particularly for heating and cooling, are highly dependent on location. If the same building were located in London, where heating degree days and cooling degree days are 2,129 and 365 respectively, the heating load becomes 75 per cent (55 per cent of cost) and cooling load 3 per cent (5.4 per cent of cost) of the total. If it were in Rome, where the heating and cooling degree days are 1,103 and 1,173 respectively, the heating load falls further to 55 per cent (33 per cent of cost) while the cooling load increases to 13.8 per cent (20 per cent of cost). This would have a significant impact on the measures to be recommended.

Table CS 1.4 Summary of energy consumption and costs for different locations

Application	Consumption kWh	% of total consumption	Cost £	% of total cost	Edinburgh
Heating	671,664	79.8	30,224	61.8	
Cooling	6,770	0.8	744	1.5	
Lighting	95,144	11.3	10,466	21.4	
Other	67,960	8.1	7,476	15.3	
Total	841,518		48,910		

Application	Consumption kWh	% of total consumption	Cost £	% of total cost	London
Heating	571,972	75.5	25738.74	55.7	
Cooling	22,873	3.0	2516.03	5.4	
Lighting	95,144	12.6	10,465.84	22.7	
Other	67,960	9.0	7475.6	16.2	
Total	757,949		46,196.21		

Application	Consumption kWh	% of total consumption	Cost £	% of total cost	Rome
Heating	296,234	55.6	13,330.53	33.9	
Cooling	73,508	13.8	8,085.88	20.5	
Lighting	95,144	17.9	10,465.84	26.6	
Other	67,960	12.8	7,475.6	19.0	
Total	532,846		39,357.85		

Case study 2

Conversion of a traditionally built dwelling to office use

This case study concerns the adaptation of an old dwelling into a small office premises for a software development company. The property is a large house about 100–150 years old, situated in a rural location in east-central Scotland. It is a traditional stone-built house with a slate roof, internal plastering, but no insulation, and a suspended timber ground floor. There are single-glazed hardwood-framed sash windows. It is heated by a very old and inefficient central heating system with large cast-iron radiators, the boiler being highly corroded and unsafe. The house was occupied by an elderly couple and has not been modernized; the electrical wiring has not been replaced for over 40 years. The water supply and drainage pipes are intact, but new lagging is required on the cold water supply. The house has been empty for two years, during which time there was a very bad winter and some of the central heating pipes burst, as did the hot water supply pipe to the taps. It has been purchased by a small financial services company that wishes to convert it to office premises. There is some land in front of the house that could be used for car parking and a yard to the rear of the premises. The usable floor area is 320 m², spread over three floors. Initially the house comprised sixteen occupied rooms, including hallways, but it will be possible to remove some internal partitions to create larger office spaces. The accommodation required comprises offices for ten staff, each with a desktop computer, and with one photocopier and two printers in total, along with a small rest room with a kettle and microwave oven. Staff toilets will be required. The working hours will be 09.00–17.30 Monday to Friday.

The first task is to establish the performance of the building as it stands. Unfortunately utility bills are not available, and the performance is to be estimated using the known characteristics of the building. Using values of thermal properties from the CIBSE Guide Part A, the U-values shown in Table CS2.1 were estimated, and the areas assessed from a survey of the building.

The sash windows are ill-fitting and allow draughts to enter, and the original flues allow some ventilation. The infiltration rate has therefore been set at two air changes per hour.

Table CS2.1 Estimated U-values for the original building fabric, case study 2

	U (W/sq.m.K)	A (sq.m)	U.A (W/K)	% fabric loss	% total heat loss
Roof	2.3	190	437	25.1	19.9
Floor	0.9	170	153	8.8	7.0
Walls	2.3	400	920	52.9	41.8
Windows	5.7	40	228	13.1	10.4
		ΣUA	1738		
		Infiltration	2.0 Ac/hr		
		Ventilation conductance	462	W/K	
		TLC	2200	W/K	

Ventilation conductance is therefore $0.22 \times 2 \times 700 = 462$ W/K. The total conductance is $1738 + 462 = 2,200$ W/K.

Using a 20-year average degree-day value for the location of 2,500 DD, the annual space heating demand is 132,000 kWh. Allowing a factor of 0.7 for intermittent heating, and 60 per cent for the boiler efficiency, this equates to an energy requirement of 154,000 kWh, or 481.25 kWh/m². This compares extremely unfavourably with the 'good' and 'typical' values of 79 and 151 quoted in the CIBSE Guide Part F. At approximately six times the 'good' value, there is clearly huge scope for improving the heating performance of the building.

Hot water and cooking energy consumption are each estimated at 3000 kWh per year. The other main supply of energy is electricity. In its original use as a dwelling, electricity was used mainly for lighting and in the central heating pump. A quick assessment is made below. Lighting three rooms at a time each with a 100 W tungsten bulb, for four hours a day throughout the year, gives annual consumption of 432 kWh/year. Energy needed for the central heating pump, assuming a 150-day heating season, and the use of 2kW for six hours a day, is 1,800 kWh (see Table CS2.2).

Table CS2.2 Estimated original energy consumption and costs

Application	kWh	% energy	Cost £	% cost
Space heating	154,000	95	6,930	93
Water heating	3,000	1.8	135	1.8
Cooking	3,000	1.8	135	1.8
Lighting	432	0.3	47.5	0.7
Other	1,800	1.1	198	2.7
Total	162,232		7,445.5	

The total electricity consumption amounts to 7 kWh/m^2 per year, far below the figures for a 'good' office. However, a direct comparison is not realistic, as far more electricity will be consumed when it is in use as an office. There will be a desktop computer for each of the ten employees, and a photocopier and two printers. The house should be rewired as a matter of course, and the wiring upgraded to allow for greater consumption. At the same time, data cabling can be installed for a local area network (LAN) and internet connections. A server will be required and can be housed in one of the refurbished outhouses; with appropriate ventilation, cooling will not be required for the server. Compact fluorescent lighting will produce a high level of savings. Table CS2.3 gives an estimate of the new electricity consumption.

Table CS2.3 Electricity consumption in the converted building

Item	Mean consumption W	Hrs/year	kWh/year
Computer × 10	2,000	2,000	4,000
Lighting 10W/sq.m	3,200	1,000	3,200
Kettle	3,000	220	660
Microwave	1,000	240	240
Photocopier	300	250	75
Printer × 2	400	250	60
		Total	8,235

Options for improving the building

Since the heating is clearly the largest consumer of energy, most of the emphasis will be placed on improving the insulation of the building and supplying the heat from more efficient plant.

Consideration should be given not only to economics but also to the practical aspects of adding insulation – in particular its thickness, since the insulation will be on the inside of the walls.

This produces the figures shown in Table CS2.5.

Table CS2.4 Proposals for the building fabric

Measure	Thickness mm	U-value W/sq.m.K
Roof insulation below rafters	140	0.25
Floor – board beneath joist	200	0.16
Walls – rigid board + plaster	100	0.35
Double glazing		2.2

Table CS2.5 Energy requirements of the altered building

	U (W/sq.m.K)	A (sq.m)	U.A (W/K)
Roof	0.25	190	47.5
Floor	0.16	170	27.2
Walls	0.35	400	140
Windows	2.2	40	88
		ΣUA	302.7
		Infiltration	1.0 Ac/hr
		Ventilation conductance	231 W/K
		TLC	2200 W/K
AHL	32,022	kWh	
Add condensing boiler			
AED	25,472	kWh	
Per sq.m	79.6	kWh/sq.m	

As there is no mains gas, the cost of gas will be slightly higher and the cost of the storage tank must be factored in, but it remains the cheapest fuel option for this location. The heating performance is in the 'good' range mentioned earlier. The natural choice of heating system would be either a traditional radiator system or underfloor heating. Underfloor heating requires a greater upheaval to the building, but as the building is empty and is to be generally refurbished this will not increase the costs significantly and can be fitted easily into the work schedule. It will take slightly longer to heat up an a cold Monday morning after the office has been closed for the weekend, but can be used effectively with high-efficiency sources of heat such as a condensing boiler or heat pump. Incorporation of optimum start/stop controls will enable maximum thermal comfort and efficiency to be achieved.

Thermostatic radiator valves should be used for local control of heating. In such a small building a building management system (BMS) is not worthwhile. A good programmable controller should be installed for the heating. Time switches or motion detectors could be used effectively with the lighting, since the wiring and controls will be renewed in any case. For domestic hot water, point-of-use heaters should be effective, since the quantity of hot water required is limited, and this obviates the necessity for renewing the hot water pipework to the taps. Table CS2.6 shows the new estimated energy consumption for the converted building, and the distribution of energy costs. The heating costs have fallen dramatically, but the electrical use has increased due to the use of computers and other office equipment.

Table CS2.6 New estimated energy consumption and distribution of energy costs

Application	k Wh	% energy	Cost £	% cost
Space heating	25,472	71.3	1,146	53.6
Water heating	2,000	5.6	90	4.2
Cooking	0	0	0	0
Lighting	3,200	9.0	352	16.4
Other	5,035	14.1	553	25.8
Total	35,707		2,141	

Table CS2.7 gives approximate costs and payback periods for the options recommended above.

Table CS2.7 Approximate costs and payback periods for the options recommended above

Element	Cost £	Annual saving £	Payback years
Walls	4,779	2,387	2.0
Roof	6,399	1,176	5.4
Floor	4,266	568	7.5
Windows	14,000	428	32.7

The boiler has not been costed as the existing one is in any case unusable; the same applies to the lighting installation.

Other options

These include the following.

Ground or air source heat pump

In the absence of mains gas, heat pumps present an efficient alternative for space heating. For a ground source heat pump (GSHP), space would be available beneath the car park or back yard for burying the coils.

Capital cost

- GSHP approx £7,000 including excavation for coils. Mean COP 4.0.
- Air source heat pump (ASHP) approx £5,000. Mean COP 3.0.
- Electrical input for GSHP = 22,415/4 = 5603 kWh. Annual heating cost £616.30.
- Electrical input for ASHP = 22,415/3 = 7471 kWh. Annual heating cost £821.81.

Compare this with £1,146 for a gas condensing boiler. The initial cost of the GSHP will be higher, since excavation is needed for the underground heating coils, probably not less than £7,000.

Payback compared with condensing boiler = 6,000/530 = 11.3 years.
Payback on ASHP = 5,000/324 = 15 years.

Although these payback periods are rather long, the maintenance required on heat pumps is lower. The optimum choice of heating system would be underfloor heating, which would allow low water temperatures and hence maximum COP to be obtained.

Renewables

Solar hot water is also a possibility. As an office, the hot water consumption will be limited to hand-washing, and will amount to no more than 6,000 kWh per year. A panel 3 m × 3 m (9 m^2) would produce about half of that amount, and a supplementary heating source would still be required to top up, and for dull days. When the cost of plumbing and connecting to the supplementary source are included, it is unlikely that a system could be sourced for less than £5,000, which would give a payback of about twenty years. The renewable heat initiative (RHI) would reduce this by only about two years.

Solar photovoltaics (PV) and biomass

Thanks to the feed-in tariffs (FIT) solar PV has become more attractive economically. A 1 kWp panel would require about 8 m^2 and would produce about 800 kWh per year. The maximum panel size for the roof would be approximately 50 m^2, producing 7 kWp. This would produce a net saving, when taking into account income from exported electricity, of £2,100, and because of the FIT the payback is reduced to only thirteen years. A biomass boiler would also be an option for such a rural location, and a detailed study would include sourcing and pricing local biomass fuel. Storage space would also be required (and is available at the back of the house).

Appendix 1
Estimating energy consumption using degree days

As heating is the main consumer of energy in buildings in the UK, it is important to acquire at least an approximate estimate of the amount of energy it consumes. In this section the concept of degree days is introduced, and a simple method of using them to make a rough estimate of annual heating energy consumption is presented. If building plans and specifications are available, then the overall thermal performance of the building can be calculated, and the annual heating energy consumption estimated using computer-based simulation models or by using the calculation methods shown below. It is a fairly simple matter to set up a spreadsheet which will make repeated calculations easier, so that the effect of making a range of changes to the building can readily be assessed. Consider a building maintained at a temperature of 20°C over a 24-hour period. The rate of energy input required to maintain that temperature depends on the indoor–outdoor temperature difference, and the thermal properties of the building, characterized by the sum of the fabric and ventilation conductances, here defined as the total loss conductance (TLC). The fabric conductance accounts for conduction heat losses through the walls, windows, doors, roof and ground floor of the building, and is calculated by multiplying the U-value of each element of the outer envelope of the building by its area.

Example

Walls of U-value 0.45 $Wm^{-2}K^{-1}$, total area 120 m^2, UA (U-value of building multiplied by area) = 0.45 × 120 = 54 WK^{-1}.

UA	(WK^{-1})
Walls	54
Roof	37.5
Windows	80
Floor	30
ΣUA	201.5

The ventilation conductance is based on the flow rate m in kg/s of cool air drawn into the building from outside, and is equal to mC where C is the specific heat in $Jkg^{-1}K^{-1}$. Inserting appropriate values of density and specific heat, this reduces to 0.33 nV where n is the ventilation rate in air changes per hour and V is the internal volume of the building in m^3.

Thus, for V = 225 and n = 2, $0.33nV = 0.33 \times 2 \times 225 = 148.5 \text{ WK}^{-1}$.

TLC = ΣUA + 0.33nV
TLC = 201.5 + 148.5 = $350.0WK^{-1}$

At any given moment the heat input required is equal to TLC × indoor–outdoor temperature difference (ΔT). It should be noted that where the air and radiant temperatures inside the building are substantially different, this simplification introduces an error in excess of 3 per cent. For many buildings the uncertainty in the input data will be greater than this. Thus the average heating demand (in W) over a day is equal to TLC × DT_{ave} where DT_{ave} is the average temperature difference over the day.

The overall heating demand in kWh is given by ΔT. TLC . 24/1000 kWh. Assume that the required indoor temperature T_{in} is 20 °C and that outdoors T_{out} is 7°C:

$\Delta T = 20 - 7 = 13K$
Total heating demand for the day = $13 \times 350 \times 0.024 = 109.2$ kWh.

The above calculations assume there are no other sources of heat within the building apart from the heating system, but this is not the case – incidental gains, such as those from the occupants (about 100 W of sensible heat for a person sitting at a desk working), solar radiation, and electrical equipment such as lights and computers all supply additional 'free' heat which raises the indoor temperature and therefore reduces the amount of heating that needs to be put in through the heating system. The rise in temperature required from the heating system is reduced by an amount equal to that provided by the 'free' gains increase to produce a *base temperature* T_b. The standard value used in the UK is 15.5°C, implying that 4.5°C of heat are provided from sources other than the heating system. The degree-day values to various base temperatures are shown in Table A1. The difference between the base temperature and the mean outdoor temperature for any given day is defined as the number of degree days for that day.

Where T_{out} is higher than T_b, zero degree days are recorded. Thus, using degree days in the above example, the effective DT is given by:

$\Delta T = T_b - T_{out} = 15.5 - 7 = 8.5K$

and the actual heating demand for the day = $8.5 \times 350 \times 0.024 = 71.4$ kWh.

To estimate the annual heating energy consumption it would only be necessary to sum the individual daily heating demand values; however, in the degree-day method the degree days for each day are summed to give an annual degree-day value (DD), whose units are degrees × days. The degree days in effect represent the cumulative temperature difference over the year. For the example above at a location in East Anglia (2,254 DD from Table A2):

$$\text{Annual heating demand} = 22{,}54 \times 350 \times 0.024 = 18{,}933 \text{ kWh.}$$

The value of T_{out} used is the mean dry-bulb temperature for that day, but if this is unavailable the average of the daily maximum and minimum temperatures may be used. Weekly and monthly degree-day totals are useful for detailed energy analysis, as shown in Chapter 3. In practice, it is not always necessary to calculate the degree days from weather data at individual locations, as they are tabulated in the CIBSE Guide A, in journals and through various internet sites for a range of base temperatures. The degree day method is a simplification, and assumes constant heating over the heating season, ignores the differences between radiant and warm air heating, and also assumes a particular level of internal and solar gains (the basis of T_b). Correction factors can be applied to allow for intermittent use. The actual amount of fuel used can be estimated by taking into account the efficiency of the boiler; the annual heating demand is divided by the fractional efficiency. In the example used, a boiler efficiency of 75 per cent would result in an annual energy use of 20,588/0.75 = 27,451 kWh.

With gas having a calorific value of 38.7 MJ/m³ this amounts to an annual gas consumption of 27,451 × 3.6/38.7 = 2553 m³ (1 kWh = 3.6 MJ).

Cooling degree days

It is also possible to calculate cooling degree days which enable the annual cooling load to be determined, but different base temperatures may be used as shown in Table A1. Using T_b =15.5°C as before, typical annual cooling degree days for London and Edinburgh would be 365 and 108 respectively.

Table A1 Monthly heating degree-day and cooling degree-hour totals to various base temperatures: London, Heathrow (1982–2002)

Base temp °C	Monthly heating degree-days (K·day) for stated base temperature											
	Jan	Feb	Mar	Apr	May	Jun	Jul	Aug	Sep	Oct	Nov	Dec
10	150	140	99	61	16	2	0	0	4	22	84	132
12	207	192	151	101	37	8	1	2	11	46	130	187
14	267	247	208	150	72	24	6	8	28	86	184	246
15.5	314	290	255	192	105	45	16	18	51	124	228	293
16	329	304	269	206	117	52	20	23	59	135	243	307
18	391	360	331	264	168	91	45	50	100	192	302	369
18.5	406	373	345	277	182	102	55	58	113	207	317	384
20	453	417	393	323	224	138	82	87	152	253	362	431

Base temp °C	Monthly cooling degree hours for stated base temperature											
	Jan	Feb	Mar	Apr	May	Jun	Jul	Aug	Sep	Oct	Nov	Dec
5	1347	1216	2166	3236	5935	7820	9965	9630	7232	5101	2507	1622
12	8	20	109	443	1626	2972	4787	4467	2454	962	158	43
18	0	0	2	32	274	635	1388	1158	308	33	0	0

Source: CIBSE Guide A. Table 2.23. Reproduced by kind permission of the Chartered Institution of Building Services Engineers.

Table A2 Mean Monthly Annual Heating Degree-day totals (Base Temperature 15.5°C) for 18 UK degree-day regions (1976–1995). Reproduced by kind permission of the Chartered Institution of Building Services Engineers

Degree-day region	Mean total degree-days (K·day)												
	Jan	Feb	Mar	Apr	May	Jun	Jul	Aug	Sep	Oct	Nov	Dec	Year
1 Thames Valley (Heathrow)	340	309	261	197	111	49	20	23	53	128	234	308	2033
2 South-eastern (Gatwick)	351	327	283	218	135	68	32	38	75	158	254	324	2255
3 Southern (Hurn)	338	312	279	222	135	70	37	42	77	157	246	311	2224
4 South-western (Plymouth)	286	286	249	198	120	58	23	26	52	123	200	253	1858
5 Severn Valley (Filton)	312	286	253	189	110	46	17	20	48	129	217	285	1835
6 Midland (Elmdon)	365	338	291	232	153	77	39	45	85	186	271	344	2425
7 W. Pennines (Ringway)	360	328	292	220	136	73	34	42	81	170	259	331	2228
8 North-western (Carlisle)	370	329	309	237	159	89	45	54	101	182	271	342	2388
9 Borders (Boulmer)	364	328	312	259	197	112	58	60	102	186	270	335	2483
10 North-eastern (Leeming)	379	339	304	235	159	83	40	46	87	182	272	345	2370
11 E. Pennines (Finningley)	371	339	294	228	150	79	39	45	82	174	266	342	2307
12 E. Anglia (Honington)	371	338	294	228	143	74	35	37	70	158	264	342	2254
13 W. Scotland (Abbotsinch)	380	336	317	240	159	93	54	64	107	206	286	358	2494
14 E. Scotland (Leuchars)	390	339	320	253	185	104	57	65	113	204	290	362	2577
15 NE Scotland (Dyce)	394	345	331	264	194	116	62	72	122	216	295	365	2668
16 Wales (Aberporth)	328	310	289	231	156	89	44	44	77	156	234	294	2161
17 N Ireland (Aldergrove)	362	321	304	234	158	88	56	56	102	189	269	330	2360
18 NW Scotland (Stornoway)	336	296	332	260	207	124	88	88	135	214	254	330	2671

Source: CIBSE Guide A. Table 2.17. Reproduced by kind permission of the Chartered Institution of Building Services Engineers.

Appendix 2

Additional data and calculations

Bills from the utility company will state the quantity of fuel used, which may be in units of kWh, cubic metres or litres, for example. It is convenient to convert everything to a common unit, such as kilowatt hours (kWh). Some conversion factors are given in Table A3.

Table A3 Fuel conversion factors

Fuel	Gross calorific value (CV)	CV in kWh	Quantity/kWh
Natural gas	38.7 MJ/cu.m	10.75 kWh/cu.m	0.09 cu.m
Medium oil	40.9 MJ/litre	11.36kWh/litre	0.088 litres
Propane	93 MJ/cu.m	25.8 kWh/cu.m	0.038cu.m
Butane	122 MJ/cu.m	33.8 kWh/cu.m	0.029 cu.m
Coal	27.4 MJ/kg	7.6kWh/kg	0.131 kg

Carbon dioxide emission factors on the basis of gross calorific value (CV) are shown in Table A4.

Table A4 Carbon dioxide emission factors on the basis of gross calorific value

Energy source	Kg CO_2/kWh
Grid electricity (UK)	0.54522
Natural gas	0.18523
LPG	0.21445
Coal	0.32227
Wood pellets	0.03895
Diesel	0.25301
Petrol	0.24176
Fuel oil	0.26592
Burning oil	0.24683

Source: courtesy of the Carbon Trust.

Discount factor calculation

The discount factors used in chapter 6 are calculated according to the following formula.

$$F(T) = \frac{1}{(1 + r)^T}$$

where:
F(T) = the factor by which future cash flow at time T years from now must be multiplied to obtain the present value.
r = fractional discount rate.
T = Time from the present in years.

Operative and dry resultant temperatures

A widely used temperature measurement for assessing the thermal comfort of humans was the dry resultant temperature, defined as

$$t_{res} = 0.5 \times t_r + 0.5 \times t_{ai}$$

where:
t_{res} is the dry resultant temperature
tr is the mean radiant temperature
t_{ai} is the dry bulb air temperature
and where the air speed is less than 0.1 m/s.

A concept more widely used now is the operative temperature, which at the low air speeds normally encountered inside buildings is equal to the dry resultant temperature.

Selected bibliography

Building Research Energy Conservation Support Unit (BRECSU). *Energy Consumption Guide ECG 019* 'Energy Use in Offices'. BRECSU, 2003.

British Standards Institution (BSI). *Energy Management Systems – Requirements with Guidance for Use.* BS EN 16001:2009.

Carbon Trust. *Energy Management Guide.* CTG054, 2011.

Chang, Wei. *Intelligent Buildings and Building Automation.* Spon, 2010.

Chartered Institution of Building Services Engineers (CIBSE). *Guide A: Environmental Design.* 2006.

CIBSE. *Testing Buildings for Air Leakage, Technical Manual TM23.* 2000.

CIBSE. *Mixed-mode Ventilation, Applications Manual AM13.* 2000.

CIBSE. *Guide B: Heating, Ventilation, Air Conditioning and Refrigeration.* 2002.

CIBSE. *Guide F: Energy Efficiency in Buildings.* 2004.

CIBSE. *Natural Ventilation in Non-Domestic Buildings, Applications Manual AM10.* 2005.

CIBSE. *Managing Your Building Services, Knowledge Series KS02.* 2005.

CIBSE. *Building Energy Metering, Technical Manual TM39.* 2006.

CIBSE. *Degree Days, Technical Manual TM41.* 2006.

CIBSE. *Refurbishment for Improved Energy Efficiency, Knowledge Series KS12.* 2007.

CIBSE. *Energy Benchmarks, Technical Manual TM46.* 2008.

CIBSE. *Guide H: Building Control Systems.* 2009.

CIBSE. *Energy Efficient Heating, Knowledge Series KS14.* 2009.

Energy Institute. *Energy Essentials. A Guide to Energy Management.* 2016.

Gevorkian, P. *Sustainable Energy Systems Engineering.* McGraw-Hill, 2006.

Levermore, G. J. *Building Energy Management Systems,* 2nd edition. Spon, 2000.

Scottish Government Building Standards, 2015.

Underwood, C. P. *HVAC Control Systems: Modelling, Analysis and Design.* Spon, 1999.

Useful organizations

Building Research Establishment (BRE), Garston, Watford, Herts WD2 7JR. 01923 664258. www.bre.co.uk

Building Services Research and Information Association (BSRIA). Old Bracknell Lane West, Bracknell, Berks. RG12 7AH. www.bsria.co.uk

Chartered Institution of Building Services Engineers (CIBSE). Delta House, 222 Balham High Road, London SW12 9BS. 0181 675 5211. www.cibse.org

The Carbon Trust administers many of the British government's energy efficiency programmes and provides advice and literature, much of it free. www.carbontrust.co.uk

The Energy Managers Association. Set up as a membership body in Feb 2012 and represents energy managers across all industries. Its purpose is to improve the position of energy management experts, establish best practice in energy management and encourage knowledge exchange. www.theema.org.uk

The Energy Saving Trust also provides energy saving advice. www.energysavingtrust.co.uk

The Energy Institute (EI). 61 New Cavendish Street, London W1G 7AR. 0171 580 7124. The professional body for those working in energy. www.energyinst.org

The Institution of Engineering and Technology (IET). www.theiet.org

The Institution of Mechanical engineers (IMechE). www.imeche.org

Index

Page numbers in *italics* denote an
illustration, **bold** indicates a table

156   *Index*